The Zone System for 35mm Photographers

A Basic Guide to Exposure Control

Carson Graves

Focal Press
Boston London

Focal Press is an imprint of Butterworth–Heinemann.

Cover design: Janis Capone
Cover photograph: Lyntha Eiler
Composition: Compset, Inc.
Printing and binding: Halliday Lithograph
Color insert and cover: Phoenix Color Corp.
All photographs by the author unless otherwise credited (see page 112)

Library of Congress Cataloging-in-Publication Data

Graves, Carson, 1947–
    The zone system for 35mm photographers.

    Includes index.
    1. Zone system (Photography) 2. 35 mm cameras.
I. Title. II. Title: Zone system for 35 mm photographers.
III. Title: Zone system for thirty-five millimeter photographers.
TR591.G73   1986   770'.28   86–2149
ISBN 0–240–51773–3

Butterworth–Heinemann
313 Washington Street
Newton, MA 02158–1626

10 9

Printed in the United States of America

Contents

8

A Working Method for 35mm Photographers

9

The Zone System and Color

Preface

One of the greatest frustrations in being a teacher of photography is watching students being limited in their efforts by a lack of understanding of how to make the photographic materials work for them. Good ideas often have trouble getting into the final form envisioned. The result is that the ideas are either compromised or that the student becomes frustrated and stops using photography creatively.

Very early in my teaching career I realized that what these students needed was a framework upon which they could build a basic understanding of how films, developers, and printing papers functioned. Such a framework should allow them to take anything that they saw or felt and translate it into practical working methods. When I discovered the zone system, which had been devised by Ansel Adams for explaining and manipulating photographic materials, I realized that it was potentially the tool that I wanted for my students.

However, the first time I tried to teach the zone system I became so hopelessly confused in the middle of my lecture that I had to excuse the class and then sit down to figure out what I had really been trying to say. The problem was that the zone system had been presented to me by the texts of the day as a recipe for advanced photographers that unaccountably mixed higher forms of mathematics and an esoteric practice called sensitometry in with the basic concepts about how film is exposed and developed. To make matters worse, most books on the subject emphasized procedures that produced a particular result and did not permit the reader to decide what was personally suitable. In short, the fundamental understanding of how photographic materials worked that the zone system provided seemed to be reserved for only an initiated few willing to wade through logarithms and to work with specialized equipment, and not for the average person interested in photography.

From my initial confusion, and through much trial and error, I discovered a way to teach the zone system to beginning students who had only their 35mm cameras and built-in light meters. I found that not only was it possible for a beginner to master the zone system, but that it was easier to learn correct technique from the beginning rather than to unlearn incorrect technique at a later date.

However, the most startling discovery about the zone system came only after I had been teaching it for awhile. As students used the system, it began to function for them as a new language might, one that defined and related ideas and feelings with the photographic process. It became the connection between their subjective vision and the technical procedures they had to perform to produce an image. In this way it made them aware of things they had only intuitively understood before and allowed them to consciously expand their photographic perception. As a result they not only became better photographers technically, but more creative as well. It is my hope in writing this book that this experience will now be shared by a greater number of people.

I would like to dedicate this book to all of the students who have so patiently listened to my explanations of the zone system and who have asked the questions that have helped me to clarify what I was trying to say.

The production of a book like this requires far more than the efforts of one person. More than anyone else Dennis Curtin has given me the support and encouragement to see this project through. Other people who have helped in many crucial ways are: David Torcoletti and Judy Canty who spent many long hours in the darkroom making prints to my specifications; Kim Van Dyke who was a willing and patient model for the exposure index test; and all the photographers who gave their images for illustrations and whose contributions are listed in the back of this book. My editor, Nancy Benjamin, was very helpful in revising my manuscript and in knowing the best places to eat on Martha's Vineyard. Also much support and help came from Sharon Fox, Klaus Schnitzer, Eliot Tarlin, and Barry Perlus. Finally, I would like to thank Arnold Gassan who was instrumental in giving me the basis for understanding the craft of photography and in forming my attitudes as a teacher.

1
Introduction

*Previous page: Photography is
a medium defined by its materials and procedures. Rather
than allowing this fact to im-
pose limitations, a photographer can work creatively with
the materials. In this image the
mechanical nature of the shutter as a measuring device for
time has created a gesture and
a form that existed only in the
mind of the photographer and
not before our eyes. In a simple
way this illustrates the photographer's need to understand
and work with the craft of photography. In the final analysis,
the only limitations that we
have are the self-imposed rules
that we place on our creativity,
not the technical nature of the
medium.*

In taking the time to read and study this book you are asking the question: Why bother learning the craft of photography beyond what is necessary to make a basic print? Many people, in fact, argue that technical proficiency in photography will not make good images, that it will only make the visual failures easier to see. In one sense this is true; craft is no substitute for vision. As some have observed, the road side of the history of photography is littered with "clear images of fuzzy concepts." However, those who choose to ignore learning craft fight the fact that the parameters, or limits, of photography are defined by the laws of optics, chemistry, and mechanics, and that these factors determine both the look and the emotional content of the final image. As long as one does not accept craft as an end unto itself, but instead approaches the connection that exists between understanding the materials and the effect they have on the whole image, then a healthy synthesis between the tools and the final product will result.

In our day to day photographic efforts the technical side of photography is constantly affecting the way our images look and feel. For example, the shutter is a mechanical device that controls the amount of light entering the camera and striking the film. Beyond that simple fact are a number of choices about how to use the shutter, each of which will produce a very different looking image. A fast shutter speed will stop moving action or render a hand-held shot sharp. A slow speed will cause movement to be seen as a blur, thus affecting the visual content of that movement and entirely changing the emotional appeal. It doesn't alter the visual experience of a painting to know how long it took to paint it, but it does matter to the viewer how long in actual time it took to make a photograph.[1]

An example of how optics affects the image is your choice of lens. We should all be conscious of how objects relate to one another in the frame. Once the print is made, all the objects will exist on a flat piece of paper in which size relationships determine the importance of how we see them. If the objects aren't in the right relationships for our photograph then we must change our position, usually by moving closer or farther away. But then the original image becomes larger or smaller in the viewfinder. It's time to change lenses to restore the image size. The point is that perspective is controlled by the camera's distance to the objects in the scene and not by the type of lens being used. Telephoto or wide-angle lenses can magnify or shrink the size of the image that we see in the viewfinder, but they do not change the relationship between objects in the photograph.

[1]Charles Harbutt, "The Shutter," in *Modern Photography* (February 1976, p. 94).

One more example relates to the chemistry of photography. Most photographers using 35mm cameras, especially those making black-and-white enlargements, are obsessed with grain. The effort to find a "grainless" film developer is like a latter-day search for the Holy Grail. Many photographers spend a great deal of money and effort in this search, and ignore the fact that exposure and development are greater factors in determining grain size than any particular combination of chemicals. In fact, the most popular "ultra" fine grain developers (such as Kodak's Microdol-X) use a chemical added to the basic developer formula that results in less sharp images and loss of shadow detail. Unless these added qualities are intentional, they may be working against the photographer's wishes.

Not taking the steps to understand our materials makes us slaves to the technology that we use. Paradoxically, photographers who pretend not to care about technique are allowing their creative

" 'Aunt' Sophie's Guest-Room Bed," by Eva Rubinstein shows how a sensitive approach to craft is used to create an expressive photograph. The glowing sensation of the light on the bed and wall was first felt by the photographer and then translated into craft decisions about film exposure and development. The photographer's feelings about this scene would not have been successfully shown in the print if it were not for an understanding of the technical side of photography.

impulses to be controlled by it. The mechanical and seemingly impersonal aspects of photography *do* affect the emotional content. Because these factors (like exposure and development) follow exact physical and chemical laws, we can predict the effect on the final image. Rather than this being a constraint, these predictable (but not unchangeable) effects can be a framework on which to hang our ideas and feelings about photographic images. Learning how the materials function is equal to freedom from any creative constraint.

This book is devoted to describing an empirical approach to understanding the zone system. This means that any theoretical information in the text can be verified by actual experience. Simple tests are outlined to provide this experience and to help in calibrating equipment. As you progress in this book, you will find that the zone system is a precise language for describing photographs. It can describe exposure and development of film in exact terms, but its real value is more than that. It is through careful description that a photographer gains an understanding of both the technical and aesthetic choices that are available whenever something is photographed. In short, the zone system becomes the first step toward answering the question about why photographs are made by encouraging photographers to be more conscious about how they make their choices.

The zone system is most fully utilized with black-and-white film, and most of the tests and examples assume its use. However, all photographic emulsions that are silver-based, including color transparency, color negative, and even the so-called "silverless" films, function according to the principles of the zone system. A chapter about color emulsions, along with special tests for them, is part of this book. Once the zone system is understood with one film, however, that knowledge can be used with all other materials.

2
What Is the Zone System?

Negative/Positive Relationship

The film that you put in your camera has a unique reaction to light. As visible light strikes the complex mixture of silver salts, bromides, and gelatin that makes up the film's emulsion, a very real but completely invisible change takes place. The exposed but undeveloped film is said to contain a *latent* image. As yet science has only theories about what produces the latent image, but the effect can be demonstrated repeatedly and consistently. The latent image is the *potential* for a visible image on film after undergoing the chemical process known as development.

The visible image is something we know more about. It consists of particles of reduced silver (reduced from a higher electrical state called a halide) that have the ability to absorb varying amounts of any light that shines through the clear film base. The more reduced silver in a given area, the more transmitted light is absorbed, and the darker the negative looks. This darkness is called *density*, and the more exposure a latent image gets, the greater the potential for density when the film is developed.

When making an enlargement from a negative, the tones that appear on the printing paper are the result of these negative densities. Whether a tone becomes a shadow, a highlight, or a middle gray on the print is controlled by how much

silver has been reduced. Shadows, the areas that we see to be darkest in our original scene and our final print, are the products of the parts of the negative that receive the least amount of light exposure and form the least density when developed. Highlights are the opposite; they appear as the lightest tones in the scene and the print, but are formed from the areas of greatest density in the negative. Only the gray mid-tone areas have approximately the same visual density in both the negative and positive print. In working with negatives, remember that dark shadow areas look the lightest (that is, transmit the most light), and that highlights will appear to be the darkest and absorb the most light.

Tonal Scale

Recorded on our negatives and prints is a range of different densities. On the print, where density is seen by reflected light rather than transmitted light, they are called *tones*. These tones vary from the maximum amount of silver that can be reduced in the emulsion (known as maximum black) to an area where only the white paper base is visible through clear gelatin and just residual amounts of silver have been left by the development and fixing process. In between these extremes lies an infinitely varied or continuous range of tones.

This concept of a continuous tone gray scale becomes a problem

Previous page: People see photographs on two levels. The most obvious level is the actual subject of the image. But also important to what we see is the way that the subject is rendered tonally. In this portrait, dark hair is surrounded by large areas of light tones that reinforce the restful feeling. One of the primary purposes of the zone system is to increase an awareness of how the tones in an image help communicate its meaning.

Negative

Positive

Photographic images consist of particles of silver that have the ability to absorb light. This absorption is called density, and the greater the exposure to light, the more density that will appear in the image. This is true of both the negative and positive images. Only the arrangement of tones are reversed. The greater the density in the negative, the lighter the corresponding tone will be in the positive print.

when we try to use it to gain control over our materials. Infinity is simply a difficult number to deal with. While realizing that we are working with continuous tone materials, it is easier if the gray scale is broken up into more manageable pieces. If we call these pieces *zones* and let each zone represent a particular area of the gray scale, we create a concept that is far more manageable than the idea of infinity. To go a step further, if we give each of these zones a name and a particular attribute as it appears on the print, then we can gain a certain amount of control over each tone. My favorite analogy to help explain this involves knowing the name of a person and the sort of power that the knowledge gives. In a crowded room we can single out that person and expect a response by calling out the name. Simply saying "Hey, you" to a crowd of people won't produce the same result. Just as the Old Testament Hebrews knew, naming a person or a concept gives one control because it implies knowledge.

Identification of Zones

In breaking up the continuous tone scale into usable pieces (zones) and learning their names and attributes, it is easiest to start at the extremes, since we have already identified them.

The tones in a photographic image continuously vary from white to black. Every area of the image represents a part of the continuous tone scale. The zone system organizes this infinite range of tones into more easily recognizable segments. Each tone of this photograph can now be identified as corresponding to a particular zone. Simplifying and naming the tones make it possible to control them in the image.

Maximum black, the greatest amount of density possible in the print, is zone I. Paper-base white is zone IX. Not only is this physical description attached to each of these zones, but there is an emotional value attached as well. A zone I tone not only is black, but it lacks any special feeling of substance, of an object existing in space. A zone IX tone is a highlight that also lacks substance of any sort. Both are simply flat, featureless, volumeless voids; one is as dark as printing paper will allow, the other is as light. This is an important consideration, as any area of a print that needs to have some sort of identity should avoid being placed in either zones I or IX.

Note on Terminology

Here, we should stop and make some special points. The zones that we are talking about are tones that exist on *black-and-white printing paper.* The language we are beginning to use to describe these tones is a language that aids us in translating the colors and fine shades of the world around us that are too subtle for any photographic material to render, into terms that the materials can understand. Learning to think in the zone system is learning to think like printing paper Photographers use the term *previsualization,* and what it means is simply that the zone system allows you to predict the final tones on the printing paper as you view the scene through your viewfinder. Later on we will learn not only to predict, but to control and to change the zones as the need or desire arises.

Another point is terminology. When referring to zones we use roman numerals (zone I, zone V, zone IX, etc.). This terminology was developed to avoid confusion with the camera functions that use arabic numerals, such as f/8, or 4 as a shutter speed meaning 1/4 of a second. Learning and using the correct terminology will be of great benefit to you later on.

Dark Shadows and Bright Highlights

After zone I and zone IX, the next zones to consider are the ones that begin to show a change from the two extremes of maximum black and paper-base white. Zone II is a noticeably lighter tone on the print than maximum black. Zone VIII is the first appearance of gray in a print as it darkens from the paper base. Neither zone shows any real detail or texture; zone II is too dense and dark, zone VIII is too light. In fact, both might be mistaken for maximum black or white respectively unless they are compared to known values of zones I and IX. Most important, though, is the feeling or emotional value given zones II and VIII. Whereas zones I and IX are flat and empty in feeling, correct zone II and VIII values give

the impression of depth and volume, the feeling of something being there but not yet defined. This is an important subtlety, because a dark and mysterious shadow in zone II will become flat and lifeless if it is rendered in zone I, and a glowing highlight in zone VIII will become empty in zone IX. Because confusion can arise in a subjective test for these values, the best way to make an evaluation is to use standard zone I and IX patches. A zone IX patch consists simply of a piece of unexposed print border; a zone I patch is a piece of paper that has been exposed to white light and normally developed to produce the darkest possible black tone. To use them, place the zone IX patch across a suspected zone VIII to see if indeed the zone VIII is a slightly darker tone. Likewise, a zone II should appear lighter compared to a standard zone I patch.

Zones with Texture and Detail

Zones III and VII are perhaps the most important zones in the process of previsualization because they are the ones that first show detail and texture in the dark and light areas of the print. Zone III might be described as the "darkest detailed shadow" and zone VII as the "lightest textured highlight." Usually, when we approach a scene to photograph it we have an idea of which areas we want to be the

darkest with detail or the lightest
with texture; therefore, zones III
and VII are the first that we
previsualize.

Very often, there are definite objects or situations that can be used
for comparative reference to these
zones. For example, zone III is the
tone of dark shadows cast by
bushes, trees, or cars on grass, dirt,
and gravel. The shadows are dark,
but need the texture of the grass or
gravel to retain a representational
appearance in the print. A person
with black hair or an animal with
black fur would normally have
these values rendered in zone III, as
dark as possible but with detail and
texture present. In short, any object
or tone that is very dark but needs
to be seen with detail on the print is
a zone III.

Zone VII also has some specific
reference points. White painted
wood on a house when it is weathered and textured, for example, is a
zone VII (smooth white siding
would be zone VIII). Unbleached,
loose-weave white fabric would be
zone VII in a print compared to
bleached, tight-weave sheets which
would be zone VIII. As with zone III
in the dark tones, zone VII is any object that is light, but needs to be
rendered with detail and texture.

The Mid-tones

The remaining zones in the scale
(IV, V, and VI) are best understood in
relationship to each other. Zone V is
a specific tone, the one provided by
Kodak for its Neutral Density Test
Card. You can buy this test card at
most photo stores; it also comes
with both *Kodak Color* and *Black
and White Dataguides.* In the zone
system, zone V is the middle of everything. Equidistant from white
and black, it is the tone that marks
the change in the gray scale from
highlights to shadows. Every zone
lighter than zone V is a highlight
and every zone darker is a shadow.
Zone V itself is neither a highlight
nor a shadow.

From zone V we get reference
points for zones IV and VI. Zone IV is
the lightest of the shadow zones.
Such tones as open shade (where
detail is meant to be readily visible)
or brown hair are standard zone IV
values. Zone VI is the darkest highlight zone. Concrete sidewalks and
Caucasian skin are standard zone
VI values. With a gray card as a reference, and these few values as a
guide, zone IV and VI can be easily
distinguished as tones one shade
darker and lighter than zone V.

Physical Equivalents for Zones

These zones are the translating devices that allow the world seen by
the eyes to be understood as a
black-and-white print. A successful
translator should know not only
what each zone looks like in the
print, but should also have some
starting points (physical equiva-

Zone I is the darkest tone that a print can be made to yield. This portrait, taken in front of a window, shows the model framed by a zone I value. Notice the lack of any detail and the feeling of a flat space.

***Zone II** is the first tone lighter
than a maximum black. The
water below the boat in this
picture is dark and without detail, but it has a sense of volume
and substance. A zone I rendering would make it flat.*

Zone III *is the darkest detailed shadow in a scene. The velvet in the display case reveals texture and form, but retains a feeling of darkness.*

Zone IV *is a medium dark tone
such as found in open shade. In
this example the shadow, although dark, has the sense of
being lighter and easier to see
into than zone III.*

Zone V *is the middle tone in the zone system. As a medium gray it is commonly the way that sunlit grass and green bushes are rendered in a print.*

Zone VI *is the first tone lighter
than middle gray. Caucasian
skin is a zone VI. Most people
are aware if the skin tones in a
portrait are not zone VI, even if
they aren't familiar with the
zone system.*

Zone VII *is the lightest textured highlight in a print. The side of this shed as well as blond hair, white clothes, and cloudy bright skies are normally zone VII values.*

The Zone System for 35mm
Photographers

Zone VIII *is the last zone with
any density in the print. This
fence as well as any object
which is smooth and very
bright will be a zone VIII. Like
zone II, this zone renders objects with a sense of volume
and substance.*

Zone IX *is the pure white of the paper base. A zone IX has no density and appears as a flat, featureless area in the print. The bright reflections on the hood and windshield of the automobile are examples of a zone IX.*

lents) taken from typical scenes that can be related to these zones. In a representational print, one in which all the tones look the way the eye normally sees them, the following physical equivalents occur in each zone:

ZONE I   (The darkest tone that a print can be made to yield) Doorways and windows opening into unlit rooms.

ZONE II   (The first noticeable tone above maximum black) Any very dark object for which a sense of space and volume is desired.

ZONE III   (The "darkest detailed shadow") Deep shadows under bushes, cars, etc., in which a sense of detail and texture is desired. Also, black hair or black fur on animals.

ZONE IV   (A medium dark tone) Average dark foliage, open shadow in landscape, recommended shadow value for portraits in sunlight, brown hair, new blue jeans.

ZONE V   (Middle gray) Kodak Neutral Test Card, average weathered wood, grass in sunlight, most black skin.

ZONE VI   (The first tone lighter than a gray card) Sandpiles, concrete sidewalks, clear north sky, Caucasian skin with sunlight falling on it, or skin photographed in overall shade or overcast light.

ZONE VII   (The lightest textured highlight) Blond hair, cloudy bright skies, white painted textured wood, average snow, white clothes.

ZONE VIII   (The last zone with any detail in it) Any very light object for which a sense of space and volume is desired. Smooth white painted wood, a piece of typing paper, a white sheet in the sunlight.

ZONE IX   (The pure white of the paper base) A specular highlight in an image, such as the reflection of the sun off of glass, chrome, or water.

These equivalents are starting points. Many others could be added, and in fact trying to do so will aid your previsualization efforts. A method given later in this book can be used to test your guesses. (See p. 90.) A final point which will be made here but discussed later is that previsualization does not just apply to a representational image. Images that make the world appear as the eye sees it are only one choice that the photographer using the zone system can make.

Relationship of F-stops and Shutter Speeds

Before we relate the zones that have just been described to the way that we use our cameras, let's review the relationship of f-stops and shutter speeds. These two functions are the means for controlling the exposure of the film. An f-stop expresses the relationship of the focal length of the lens (hence "f" or "focal" stop) to the size of the hole (aperture) created by the adjustable diaphragm. The f-number is the denominator of a fraction which states the actual size of the aperture relative to the focal length. A 100mm focal-length lens at f/2 has an aperture diameter of ½ the focal length or 50mm. Thinking of f-stops this way will help clear up two usual points of confusion. First, as f-stops get larger in numerical value (2, 2.8, 4, etc.), the actual size of the aperture gets smaller since ½ is larger than ¼, and so on. Second, the same f-stops on different focal length lenses have different diameter apertures. For example, a 50mm lens at f/2 has an aperture diameter of only 25mm compared to twice that on a 100mm lens at f/2. In spite of the different size openings, both the 100mm lens and the 50mm lens will give an equal amount of exposure to a piece of film at f/2. This principle is true for all lenses of any focal length.

A change in an f-stop will change the exposure by a factor of two. Changing from f/5.6 to f/8 will cut the exposure in half. Changing from f/5.6 to f/4 will double it. This is why the standard progression of f-numbers seems to make no sense. The fractions expressed by the f-numbers are calculated to give apertures that relate to each other by this factor of two in terms of ex-

posure, not mathematical logic. The best way to deal with this problem is to simply memorize the standard sequence so that you can avoid f-stops that don't change the exposure by this factor. The sequence goes: f/1.2 (the largest aperture normally found, lets in the most light), f/1.4, f/2, f/2.8, f/4, f/5.6, f/8, f/11, f/16, f/22, f/32, f/45. Knowing the sequence allows you to make changes in exposure quickly while looking through the camera and still know what the aperture will be. It also prevents the use of a nonstandard f-stop. Many camera manufacturers put a nonstandard f-stop (such as f/1.8, f/1.9, and f/3.5) on the maximum opening of their lenses. These do *not* express an exposure change by a factor of two and should be avoided in order to keep control of exposure.

Like f-stops, the numbers on the shutter speed dial of a camera are fractions of the actual speed; for example, 2 is ½ of a second, 60 is 1/60 of a second, and so on. From one number to the next, the exposure time is changed by a factor of two. The shutter speed marked 250 is twice as fast as 125; 30 is double the exposure time of 60.

Law of Reciprocity

Since f-stops and shutter speeds both affect exposure by a factor of two (even though they differ in the way they affect the look of the image) they are said to have a one-to-

*Using different combinations
of f-stops and shutter speeds
both these images were shot at
the same exposure. This is an
example of the law of reciprocity. What f-stop and shutter
speed you choose to use can affect the way the image looks.
The left-hand photograph was
shot with a fast shutter speed to
stop the action. Since a large
aperture (f-stop) had to be
used, the image has very little
depth of field and the background is out of focus. The
photograph on the right was
shot with a small aperture and
the depth of field has increased
to make the background sharp.
The moving figure is blurred,
however, because the small
f-stop required a slow shutter
speed.*

**Fast Shutter Speed, Shallow Depth
of Field**

**Slow Shutter Speed, Great Depth of
Field**

one or *reciprocal* relationship. This
fact, known as the Law of Reciprocity, means that in a given lighting
situation a change can be made in
either the shutter speed or the
f-stop by making a corresponding
opposite change in the other and
still keep the same exposure. Thus,
we can choose to affect motion or
depth of field for any image we
make.

It is interesting at this time to
de-mystify an often misused term
in photography, that of *reciprocity
failure.* All reciprocity failure means
is that in certain extreme situations
when the shutter speed is either
very fast (above 1/10,000 of a
second) or very slow (slower than 1
second) the one-to-one relation-

ship between shutter speeds and
f-stops breaks down, or, the law of
reciprocity fails. In this case a
greater-than-predicted exposure is
required to obtain a desired film
density. In actual practice, the
35mm camera user will rarely encounter reciprocity failure.

Relation of Zones to
Camera Controls

To complete our description of the
zone system, we need to relate the
two different ideas we have learned
in this chapter: first, how the print
is made up of various tones called
zones and second, how the camera's f-stops and shutter speeds

function as exposure controls. The link between the two concepts occurs in the film's density where a change in exposure of one f-stop (or shutter speed) will change the tones in the print made from that negative by one zone. In other words: *Each zone in the zone system is related to the other zones by an exposure factor of one stop for each zone.* This is why there are nine zones instead of eight or ten. It takes nine stops to change film density from what will be rendered as maximum black to what will print as paper-base white on normal photographic enlarging paper.

Placement

The implication of this idea is that we have the final say over what tones occur in our prints. For example, if we have a zone V in a scene and want it to be a zone VI instead, we simply make a change in exposure of one stop. Because we are going from a darker zone to a lighter zone, more exposure is needed to increase the film density. The final exposure to produce a zone VI tone is one stop *more* than what was needed for the original zone V.

This sort of manipulation is called *placement,* and it enables a photographer to choose a tone in the original scene and place it in any zone that is needed for the final print. How this is done and the limitations that the materials put on

this manipulation are the subjects of the next chapter, "Measuring Light." The important thing to remember now is that zones and the camera's exposure controls (f-stops and shutter speeds) are directly related. Any change in exposure will change the zones that appear in the final print and any change in zones (through previsualization) will require a change in exposure.

3
Measuring Light

What Light Is—and Isn't

In 1803 an Englishman named Thomas Young proved through an elegantly simple and easily repeatable experiment that light was made up of continuous waves much like ocean waves or the way a string vibrates when it is stretched between two points and shaken.

In 1905 Albert Einstein developed a theory that light was composed of tiny discrete particles, which he called photons. This theory is one of the building blocks of a study of physics called quantum mechanics, and it won Einstein a Nobel Prize in 1921.

Neither theory, however, can disprove the other. Whether light is made up of waves like ripples on a pond or tiny particles shot out like bullets from a gun is a paradox that physicists can't yet resolve. One theory has even gone so far as to suggest (with appropriate mathematical proof) that light is a living organism that can make choices and respond to situations.[1]

Reflected Light

Fortunately, photographers don't have to know what light is made of in order to control it. The important thing to know is that light is re-

flected off everything we see, colors as well as shades of gray. In fact, what we think of as a physical presence is nothing more than the brain's interpretation of light that has bounced off of objects and entered our eyes.

Photographic film also "sees" reflected light. Density, in fact, is a product of reflected light. The surface of every object that we can see reflects a certain amount of the light that falls on it. A white sheet of typing paper (zone VIII—remember?) might reflect 90% of the light, whereas a zone II piece of black velvet would reflect only 5 to 6% of the same light. The percentage of overall (ambient) light that is reflected determines how light or dark an object appears to us. As a note of explanation, an object that reflected 0% of the ambient light wouldn't be visible and an object that reflected 100% would become a light source itself.

Colors, too, are a product of the percentage of reflected light. Whereas gray tones reflect the entire light spectrum evenly, colors reflect a percentage of only a part of the spectrum. Red objects, for example, reflect only the longer wavelengths while blue objects the shorter wavelengths. Any dark color reflects a small percentage of that wavelength and any light color a greater percentage.

Measuring reflected light and being able to relate it to previsualization is an important part of using the zone system.

Previous page: Being able to measure the light reflectance in a complex scene is the first step in controlling the tones of that scene. In this image the highly reflective snow required special placement to be rendered as a zone VII. Allowing the meter to "average" this scene would have produced an underexposure.

---

[1] *The Dancing Wu Li Masters: An Overview of the New Physics*, by Gary Zukav (New York: Bantam Books, 1979).

The Light Meter

The photographer can use two tools to measure reflected light. The most sensitive and widely used tool is the photographer's own eye. Unfortunately, our eyes are never constant; they are always adjusting for variations in light intensity so we can never get an accurate comparison between different objects.

The photographer's second best tool is the light meter. Most handheld and all in-camera light meters read reflected light. Because they are electrical/mechanical devices, these meters are calibrated to measure light against a constant value. This allows them to distinguish between objects that have different reflectances. A reflected light meter will read zone IV brown hair and a zone VII white building differently in the same light: each reflects a different percentage of the light.

Some light meters, instead of reading reflected light, read the amount of light falling on a scene. Called incident meters, nearly all are hand-held devices distinguished by a white plastic dome or cone covering the light-sensitive cell (many reflected light meters have incident attachments). The primary problem with incident meters is that they give no indication of the relative reflectance of different tones. The white building in the example above would give the same reading as the brown hair if they were both measured with an incident meter.

Medium Gray

As mentioned earlier, every light meter must have a standard against which to measure light. This constant value built into the meter is a gray tone that reflects 18% of the light that falls on it. This means that any reading that a light meter gives a photographer (called an *indicated meter reading*) will render an average density on film equivalent to an 18% reflectance gray tone when printed. This will be true even if the tone metered is a zone III detailed shadow or zone VII white snow. The indicated meter reading produces a film density that makes the dark shadow appear lighter than previsualized and the white snow darker. Both will be made into a medium gray tone. Only when a scene in which there are approximately equal areas of dark shadows, bright highlights, and middle tones is metered will the various reflectance values average out to 18% or medium gray. Since most photographic scenes have about the same amount of dark and light tones in them, they average about 18% re-

Every tone has a specific reflectance value that determines how it will look relative to the other tones in a scene. The piece of white paper (zone VIII) reflects about 90% of the light striking it, whereas the dark cloth reflects only about 5–6% of the same light and is seen as a zone II. For comparison, a gray card (zone V) reflecting exactly 18% of the light is shown between the two. Caucasian skin tone reflects about 30–40% of the light.

flectance. Usually indicated meter readings of these scenes give adequate exposures.

The problem with average overall meter readings is the word *average.* These readings will not be correct when the tones in the scene do not reflect a combined 18% of the ambient light. A photographer relying on only average readings either accepts poor exposure (and print quality) for these scenes or learns to avoid them altogether. This can make the word *average* as much a judgment of the image as it is a measurement of light.

The consequences might be this: In a scene in which mostly dark tones are previsualized, an "average" reading will make it too light and spoil the feeling that a detailed zone III shadow will produce. The same will hold true for a scene composed of primarily light tones. The delicacy of a correct zone VII textured highlight will be destroyed by making it appear too dark and gray. As you begin to push your vision into picture-taking situations where the tones do not have an overall 18% reflectance, you will no longer be content to accept "average" as either a lighting condition or a description of your pictures.

The solution to this problem lies in relating the 18% reflectance value to the zone system scale that we discussed in Chapter 2. There is, in fact, a direct connection: the 18% reflectance tone that the light meter is calibrated to read is a medium gray identical to zone V.

Exposure Corrections for Placement

By relating an indicated meter reading of zone V to the zone system scale, a correct exposure (one producing on film the densities needed to print a previsualized tone) can easily be determined for any scene. Since the relationship between each zone in the gray scale is an exposure factor of two (one f-stop or shutter speed change), then knowing that the light meter always starts out at zone V means that you can correct an exposure to obtain any zone you want.

For example, to get the correct exposure for a previsualized zone IV head of dark brown hair, an indicated meter reading of 1/60 sec at f/8 must be closed down to 1/60 sec at f/11 (or 1/125 sec at f/8 depending on whether you want greater depth of field or stopped action). The reason is that at the indicated meter reading of 1/60 sec at f/8 the dark brown hair previsualized at zone IV would be rendered as medium gray (zone V). One stop less exposure creates a density on the film that will print as zone IV.

As we have mentioned, this exposure correction is called *placement.* Every zone from I to IX can be placed by counting the number of zones that it is from zone V and opening for highlights or closing for shadows one f-stop for each zone. A zone III shadow is placed by taking the indicated meter reading and stopping down two stops. Con-

Shadow Left in Zone V

*Placement is necessary be-
cause all light meter readings
start us at zone V. If a tone is
previsualized in any zone
other than zone V, then a
change must be made in the
exposure. In the three examples here and on pp. 30–31, a
zone is shown, first unplaced
and left in zone V, and then
placed into the previsualized
zone. In each case the placed
exposure was changed from
the indicated meter reading by
one stop for each zone previsualized from zone V.
In these two images, the
shadow of the tree is previsualized in zone III. The indicated
meter reading is stopped down
two stops.*

Shadow Placed in Zone III

*The Caucasian skin tone is pre-
visualized in zone VI. Indicated
meter reading is opened up
one stop.*

Face Left in Zone V

Face Placed in Zone VI

Snow Left in Zone V

Snow Placed in Zone VII

The snow is previsualized in zone VII. Indicated meter reading is opened up two stops.

versely, a zone VIII highlight is
placed by opening up three stops
from the indicated meter reading.

Placement, then, is the method
by which you are able to relate the
way your eye sees reflected light
(previsualization!) to the way the
light meter and film see reflected
light. A systems analyst would call
placement the point of interface between man and machine. The zone
system simply says that placement—the two-step process of 1)
previsualizing, and 2) adjusting the
indicated meter reading to the previsualized zone—allows you complete control over the tones created
by your exposures.

How to Use the Camera's Built-in Light Meter

To successfully place an exposure
requires knowing how to make accurate light meter readings. The
light meter built into most 35mm
cameras works very well as long as
you keep certain facts in mind. All
of these meters average out the
tones found in the viewfinder to
zone V medium gray. This includes
"center weighted" metering sys-

tems which reduce only the area in
the viewfinder that most of the
reading is made from. To use these
meters to place exposures, a previsualized tone must be isolated so
that the meter reads only the light
reflected from that area. This means
walking up to the tone so that it
completely fills the viewfinder, and
adjusting the camera's f-stops and
shutter speeds so that the meter
indicates the correct exposure in
the usual manner. Next, adjust the
indicated meter reading to place
the exposure into the previsualized
zone. Once that is done, step back
to frame the entire scene, and, ig-
noring any further changes in what
the meter indicates, take the
picture.

One caution in this procedure is
that the previsualized areas of the
scene must be large enough to fill
the viewfinder without forcing the
camera so close that the lens (or the
photographer) casts a shadow on
the area and causes a false reading.
With a normal (50mm) lens, a good
size for the previsualized area is a
minimum of 8 x 10 inches. If you are
using a wide-angle lens and it prevents isolating a single tone, then
substitute a longer focal length lens
to take the reading, and switch back
to the wide-angle for the exposure.
It is not necessary for the area being
metered to be in focus to measure
the light reflected from it. In cases
where the previsualized areas are
too small or too far away to be
metered accurately, a substitute
usually can be found, a process we
will discuss later. (See pp. 89-90.)

*Light meters built into 35mm
cameras read the light reflected from all the objects that
appear in the viewfinder. To accurately read the light reflected from a single tone it is
necessary to get close enough
to it so that the area of that tone
completely fills the viewfinder
(focus is not important). The
settings that the meter rec-
ommends for that tone is the
indicated meter reading and
will then have to be placed
according to the
previsualization.*

4
Understanding Contrast

Contrast and Previsualization

Until this point we have assumed that the amount of light reflectance always doubles or halves (a change of one stop) for each zone that we previsualize. This would mean that a zone IV shadow reflects half the amount of light that a zone V mid-tone reflects, and a zone VII textured highlight reflects four times (two stops) more light than the zone V. If light always acted in this manner, the zone system would end with previsualization and placement. But, just as quantum theory in physics speculates, light seems to think and act on its own, and more often than not measured reflectance values are greater or less than one stop per zone.

The problem is that previsualization is only our mental image of how we see zones and how we want the final print to look. It may or may not fit the actual reflectance when measured by a light meter. Just because we visualize two tones as being four zones apart (such as a zone III and a zone VII) does not mean that the zone VII will reflect four stops more light. The *intensity* of the ambient light, rather than our previsualization, affects what the meter reads.

Previous page: Light is an almost magical substance that makes certain photographs come alive. There is no secret to understanding its role in an image such as this. This image is simply a combination of previsualization and careful measurement of light.

Light Intensity and Contrast

To understand this, imagine an arbitrary unit of light intensity called a "lumen."[1] Suppose 10 lumens of light are falling on a scene in which a zone III (darkest detailed) shadow and a zone VIII (bright with little or no texture) highlight are previsualized. Imagine also that the zone III reflects 10% of the light falling on it, and the zone VIII 90%, which is approximately what they would reflect in a typical situation. With 10 lumens of ambient light, the zone III would reflect 1 lumen and the zone VIII 9 lumens. This makes the range of contrast for the scene 8 lumens, since contrast is the difference between the highlight reflectance and the shadow reflectance.

Now, imagine an increase in the amount of ambient light in the scene to 100 lumens. Our previsualization of the scene hasn't changed; we still want the final print to look the way we first saw it. The percentage of reflectance stays the same too since the objects are still made of the same substance. The only change is in the intensity of the light. But, with this change the zone III now reflects 10 lumens

[1]A lumen actually exists as a measurement of light intensity although in a slightly different context. It is the rate at which light falls on a certain measured area and is usually expressed in units of time, such as "lumen-seconds." That definition is not needed here.

Light is a constantly changing substance with the ability to alter the contrast of a scene. In the photographs illustrated here, the only difference is the intensity of the light, yet there is a profound difference in what we see. We must be sensitive to the quality of the light in our photographs and constantly test our previsualization with the light meter.

Low Intensity Light

High Intensity Light

of light and the zone VIII 90 lumens. This makes the range of contrast 80 lumens, a considerable increase from the first example. So, instead of contrast being a constant, it is actually relative to the overall intensity of light. We should always expect contrast to be different on a bright sunny day than it would be at dusk or when the light is dim.

Kinds of Contrast

Three terms describe the different possibilities for contrast in a photographic scene. *Normal contrast* is when the light reflected equals one stop of exposure difference for each zone previsualized between highlights and shadows. *Low contrast* is when reflectance values are less than one stop for each previsualized zone. *High contrast* is when reflectance is greater than one stop per zone. Each of these contrast situations has a certain visual appearance and a way to relate that appearance to the zone system by using a light meter.

NORMAL CONTRAST
Normal contrast in a typical scene has a visual appearance of an even range of tones, not very dark in the shadows, not too brilliant in the highlights. The shadows cast by directional light appear fully defined but not too harsh. There is an overall pleasing, comfortable feeling from the light. Usually, normal contrast scenes are found in the open shade of the north side of a building

on a sunny day, or out in the open on an overcast day between the hours of 10 A.M. and 3 P.M. Remember, however, that appearance to the eye is subjective when comparing reflectance values. Although experience and observation are useful in giving a general idea of the contrast of a scene, only careful measurements with a light meter will be a true indication.

In using the light meter to determine the contrast of a scene, the indicated meter readings from previsualized highlight zones and previsualized shadow zones are compared. It is easiest to express the comparison of these indicated meter readings as a difference in f-stops. In a normal contrast scene the number of stops difference between two zones should equal the number of zones. For example, in a scene in which a zone III and a zone VIII are previsualized, there should be a five stop difference in the indicated meter readings (subtract zone III from zone VIII to get a five zone difference). This difference in stops is always relative to the previsualization: normal contrast can be indicated by two stops of exposure difference when zones IV and VI are previsualized, or four stops of difference when the previsualized zones are III and VII. Remember that the comparison must be made between a shadow zone (I-IV) and a highlight zone (VI-IX). The differences within the shadow zones or the highlight zones only will not be enough to give an accurate determination of contrast.

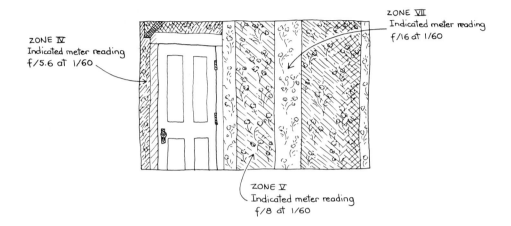

ZONE VII
Indicated meter reading
f/16 at 1/60

ZONE IV
Indicated meter reading
f/5.6 at 1/60

ZONE V
Indicated meter reading
f/8 at 1/60

Normal Contrast: *Normal contrast occurs in an image when the number of f-stops between the indicated meter readings of the most important highlight and shadow zones equals the number of zones. In this image, the zone IV previsualized for the wall on the left had an indicated meter reading of f/5.6 at 1/60, and the zone VII wall in the middle of the frame had an indicated meter reading of f/16 at 1/60. These two tones are three stops apart in reflected light values and three zones apart in previsualization. Hence, normal contrast.*

LOW CONTRAST

Low contrast scenes occur when reflectance values are less than one stop for each previsualized zone. To the eye a low contrast scene is one of even, nondirectional light. Shadows cast by objects are either soft and indistinct or nonexistent. Low levels of light intensity most often produce a low contrast scene, but not always. A notable exception would be a landscape covered by snow or a light sandy beach on a sunny day. Because light sand and snow are highly reflective in all directions, a sunny day will simply "bounce" some of that excess light into the shadows, "filling" them in. The result is a low contrast scene.

As in normal contrast scenes, the only way to make more than an educated guess about low contrast is by measurement with a light meter. The reflectance of light from the previsualized highlights and shadows in a low contrast scene will create less of a difference in stops than there are zones. From a scene with a zone III and a zone VIII there would be only a four stop or less difference when the indicated meter readings from each zone are compared. Here, too, the range of stops indicating low contrast is relative to previsualization. Zones III and VII would reflect three or fewer stops of exposure difference; zones IV and VI only one stop difference.

HIGH CONTRAST

A high contrast situation, one in which there are more measured stops of exposure difference than zones previsualized, is characterized by strong intense light, casting deep, sharply defined shadows. Most bright sunlit scenes are high contrast, except for landscape scenes of snow or sandy beaches, as noted above. So, too, are scenes lit by bright artificial lights such as stage productions. Any scene in which the range of tones go from very bright in the highlights to very dark in the shadows will usually be high contrast.

A high contrast scene is defined by a greater number of stops between meter readings than there are zones between the previsualized highlights and shadows. Zones III and VIII will measure six or more stops difference, or zones IV and VII at least five stops when there is a high contrast situation. As in the other types of contrast situations, final determination must be made by comparing the indicated meter readings to the zones that are previsualized.

Specific Contrast Terminology

A shorthand notation has been developed to describe different kinds of contrast. Normal contrast is referred to simply as "N." Low contrast is called "N+" because there are more previsualized zones than metered stops. If there is one extra zone (for example, a scene in which there are three stops difference be-

ZONE III
Indicated meter reading
f/5.6 at 1/125

ZONE VII
Indicated meter reading
f/16 at 1/125

Low Contrast: *When meter readings show fewer stops of exposure difference than previsualized zones, we have low contrast. Here the black-dressed mannequin (zone III) and the white-dressed mannequin (zone VII) read f/5.6 at 1/125 and f/16 at 1/125, respectively. The difference of four zones in previsualization but only three stops in indicated meter readings means N + 1 contrast. This image is printed to show low contrast; on p. 79 you can see what it looks like printed normally.*

High Contrast: *High contrast in a scene is indicated when the meter readings show that light reflectance is greater than the previsualization. The figure shadowed by the window light is an example. The face was previsualized as zone IV and the sunlit wall as zone VIII. Indicated meter readings were f/2.8 at 1/125 for the zone IV and f/11 at 1/250 for the zone VIII. With five stops of exposure difference but only four zones, the scene is N − 1 contrast. This print illustrates high contrast. On page 80 is an example of the image correctly printed.*

ZONE IV
Indicated meter reading
f/2.8 at 1/125

ZONE VIII
Indicated meter reading
f/11 at 1/250

tween zones III and VII) it is called N+1. When there are two extra zones (such as three stops between zones III and VIII) the term is N+2, and so on. High contrast is called "N−" since there are fewer zones than stops. One less zone (five stops, for example, between zones III and VII) is an N−1, and two fewer zones (seven stops between a III and an VIII) would be an N−2.

These terms will also be used to refer to development times for film in Chapter 5. "N" refers to a development time for normal contrast scenes, "N+" refers to an increase in development time to compensate for low contrast, and "N−" refers to a decrease in development time to compensate for high contrast.

Learning this terminology will be a convenience, but more important, using it enables you to specify the exact kind of contrast in a scene. As part of your vocabulary, these terms will help you to think more clearly about what you are photographing.

The Importance of Previsualization in Determining Contrast

Contrast is relative to previsualization and the measurement of reflected light by a light meter. A normal contrast scene occurs only when the previsualization of a highlight and a shadow zone fits the light meter's measurement of one stop for each zone. This is inde-

pendent of any type of subject matter or lighting situation. The same is true for high and low contrast. Although general observations can be made, the ultimate determination is the relationship of previsualization to light meter measurement.

Through the exercise of your previsualization, you have the means to control the tones of any photographic scene that you might encounter. This makes it your responsibility for the final look of the photograph. Once equipment and procedural error are eliminated, any failure of the image to produce the desired result is one of previsualization. If a shadow is too dark or a highlight too light, then you must rethink how you mentally saw the photograph before exposure.

Previsualization failure will occur often as you learn the zone system. You may make prints where tones appear as previsualized, but the image does not possess your feelings at the time of previsualization. This happens because the translation of the mental equivalent of each zone into the actual print is a process that can be fully understood only through practice. When failure does happen, you must ask yourself whether the shadows and highlights were previsualized in the correct zones. Perhaps you might have obtained a better effect if you had placed a shadow in zone III instead of zone IV, giving the maximum effect of darkness while still retaining detail rather than a more open shadow tone. This is a differ-

*Even when light reflectance
remains the same, different
previsualizations can create
different contrast scenes. In (a),
the barn wall on the right was
previsualized as zone IV and
the picket fence as zone VI. The
indicated meter readings of
these tones were only one stop
apart, so an N+1 development
was given the negative. In (b),
the model's sweater was previsualized as zone IV and the
white notebook as VIII. Here
the readings of these tones
were five stops apart and an
N−1 development was needed.
What changed (aside from the
addition of the model) was the
previsualization, not the light
intensity, nor the reflectance
values of the objects in the
scene.*

(a) N+1 Scene

(b) N−1 Scene

ence of one stop of exposure, or exactly half the amount of light that would be used in a zone IV previsualization. Or, perhaps a highlight should have been previsualized as VII instead of an VIII, utilizing the detail of the highlight area instead of just a rendering of an object existing but having no definite appearance. This will change the contrast of the scene. These factors make a difference in how the print looks, and these are precisely the factors that you now have direct control over.

5
Film Development: Making the Zone System Work

Historical Background

In the nineteenth century photographers had to make their own film emulsions. Daguerreotypes, wet plates, even the first "dry" plates were all to a certain degree manufactured by the individual, often working in the field just before exposing. This did not allow for any standardization; each photographer had a jealously guarded formula worked out through trial and error. Exposure was intuitive and most photographers felt that through development they could alter the negative at will, correcting for mistakes in exposure and changing the relationships of the tones, independent of any other factor. Because there was no standard to observe, no one really understood the connection between exposure, development, and the final image.

At the end of the nineteenth century when consistent film emulsions began to be manufactured (George Eastman established his reputation making reliable dry plates), the superstitions surrounding film development were forced to change. Two amateur photographers and scientists named Charles Driffield and Ferdinand Hurter became interested in these new dry plates. They turned an old sewing machine and the light from a single candle into an experimental apparatus that showed for the first time that photographic materials gave consistent and predictable re-

sults.[1] What they discovered may seem familiar to us now, but it shocked the photographic world of the time, and it is worth repeating here because it forms the basic principle of the zone system.

Through careful testing, Hurter and Driffield discovered the true relationship between exposure and development as they related to film density. Density was first of all directly related to exposure. Contrary to what photographers believed at the time, no amount of development could correct for an under- or overexposed negative. Hurter and Driffield also discovered another interesting and for us a crucial fact. Changes in development, whether in time, temperature, chemical concentration, or even patterns of agitation, cause changes in film density, too. These development changes, however, affect density differently than changes in exposure. The highlights, areas of the negative that have received the greatest exposure, develop density more rapidly than the shadows. In fact, the lower shadow zones (III and below) that have received the least exposure in the negative change density hardly at all no matter how the development changes. These facts

---

[1] The history of photography is a fascinating story. Two well-written and informative books on the subject are: Arnold Gassan's *A Chronology of Photography* (Rochester, NY: Light Impressions Corp., 1982) and Beaumont Newhall's *The History of Photography* (New York: The Museum of Modern Art, 1964).

Previous page: The bright sunlight in this scene created a high contrast situation. Without losing the sense of light, the densities of the negative were brought under control and made printable by a less-than-normal development time.

are expressed in the zone system by the phrase "expose for the shadows and develop for the highlights." This simple statement is an acknowledgment that the shadow zone densities are affected primarily by exposure whereas the highlight densities can be altered by the act of changing development.

The Effect of Development on Shadows

Hurter and Driffield's discovery leads us to the final understanding of how to control the tones of our negatives. Exposure placement controls the primary shadow areas of our scene. Development doesn't have much effect on these densities. A graph will illustrate why, but before looking at the graph, a few words are in order about graphs in general.

A graph is a visual way of simplifying complex ideas. It is an abstraction, in this case an abstraction of what happens to film densities during development. It eliminates a lot of extra visual material that would distract us if we looked directly at the negative. In our graph we are showing the effect of different development times on film density. The horizontal axis, reading from left to right, represents increasing development time, and the vertical axis as it goes higher shows increased film density. Where the two meet is zero for both.

What we see is essentially a flat line, or what in geometry is called a curve with a small slope. This tells us that a shadow zone on film is not changed much by changes in development. As development time lengthens along the horizontal axis, the density does not move up the vertical axis. In fact, the only way to change the location of the curve is to change exposure. The dotted lines represent an increase or decrease in exposure, but again development is not a factor in the change. (See Figure 5-1.)

FIGURE 5-1.

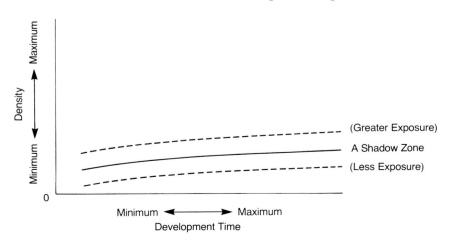

Shadows are the zones most affected by changes in the exposure of the negative. These three negatives illustrate the information in the graph (Figure 5-1). The correct exposure (b) renders the zone III pants with adequate detail whereas the underexposure (a) does not. An overexposure such as (c) has as much detail in the shadows as (b) but also has extra density, which will show up in the print as more noticeable grain.

(a) Underexposed

(b) Correctly Exposed

(c) Overexposed

The Effect of Development on Highlights

A graph that shows what happens to highlight densities during development would look much different. The slope of the curve is greatly increased. Starting at about the same density as the shadow zone, the density of the highlight rapidly increases as development proceeds. What we see in this graph is that without changing exposure, a highlight can be changed in density by varying development. (See Figure 5-2.)

How Development Controls Contrast

These two points, that shadow densities are controlled by exposure and highlight densities are controlled by development, are the cornerstones of the zone system. "Expose for the shadows," as we have seen, refers to placement. "Develop for the highlights" refers to the amount of contrast wanted in the negative.

When we combine the graph of the shadow densities with the graph of the highlight densities, the result shows that development time determines how far apart the two will be. (See Figure 5-3.) This is exactly what we defined in Chapter 4 as contrast. At one point during development, both the highlight and shadow densities will produce the tones on the print that were

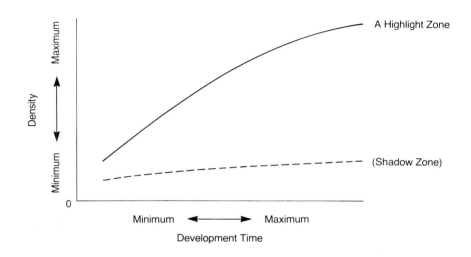

FIGURE 5-2.

previsualized. Before that time, the tones will be too close together and the image will be low in contrast, or flat. Longer than that time and there will be too much separation; the result will be high contrast. Because there are many possible ways to previsualize the contrast of a scene, there can be no standard development time for film. Each time we previsualize a scene to be photo-graphed, whether it is N, N+1, or N−2, we are specifying a different development time, one that will develop the highlights to the density that fits our needs for the image. In black-and-white each film and developer combination is different, so that the only way to discover the actual time for a specific contrast is through testing. This is the subject of Chapter 7.

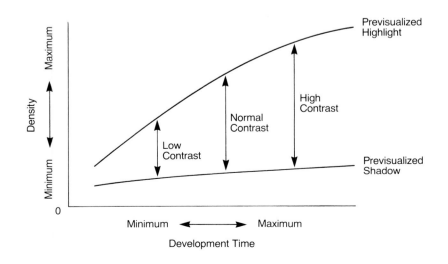

FIGURE 5-3.

*As development time in-
creases, so does contrast.
There is only one point during
development that the densities
of the negative will produce the
previsualized tones on a print.
Print (a) illustrates underdevelopment. When the gray card
is correctly printed, the white
shirt (previsualized as zone
VII) appears too dark and the
black pants (previsualized as
zone III) appear too light. Print
(c) has the opposite problem.
The negative was over-
developed, causing the shirt
to appear lighter than previsualized and the pants darker.
The only print that fits the previsualization is (b), the one
made from a correctly de-
veloped negative.*

(a) Underdeveloped

(b) Correctly Developed

(c) Overdeveloped

Technical Control Versus Creativity

In some ways nineteenth century
photographers were right about the
final step in photographic creativity
being in the development process.
But they were wrong in assuming
that changing the image with development could be subject to the
whims of the photographer. Hurter
and Driffield proved that density
changes and contrast were a product of exposure first, and then development in a completely predictable chemical process. This was a
great blow to photographic artists
of the day. It reduced to simple
cause and effect mechanism what
to them had been artistic preroga-
tives. The photograph now seemed
to rely too much on mechanical devices and chemical baths rather
than the kind of pure and direct
communication that a painter has
with the canvas to be considered an
art form.

A champion of art photography
in the nineteenth century, a doctor
named Peter Henry Emerson, even
went so far as to publicly denounce
the idea that photography could
any longer be considered art when
he heard about Hurter and Driffield's experiments. Interestingly,
however, photography didn't
change as a result of Emerson's dis-
satisfaction. In fact, even Emerson
continued to make the same sensitive images of rural England that he
had before renouncing
photography.

An important lesson is to be learned from this. Before Hurter and Driffield's discovery, photography had been a creative medium only because photographers thought they could randomly apply technique to produce a desired result. Since that time has come the realization that only through the understanding and application of careful and predictable craft can photographers be free to visually express their feelings in harmony with the materials. As is true with any expressive form, the photographic materials require a certain discipline that has to be understood before it can be effectively used.

Making the Zone System Work

The pieces of the zone system puzzle are now complete. Putting them together and making them work is a relatively simple task. It involves, first, selecting a scene and deciding through previsualization how you want the final print to look. The next step is to meter the various previsualized areas. And, finally, use the indicated meter readings to determine placement for exposure and contrast for development time. Assuming that there are no technical flaws in the materials, equipment, or your testing procedures, the negative that you produce will print on #2 (normal contrast) paper without any difficult printing ma-

nipulation. This is one of the prime functions of the zone system, to simplify and take the guesswork out of the technical process. Let's break this process down and discuss each of the parts individually.

SUGGESTIONS TO AID PREVISUALIZATION
We know that previsualization means mentally sizing up an image before you make the exposure. It is a way to see the scene in the same tones of black-and-white that the film and enlarging paper do. But, given all the choices available, confusion can arise as to which zones are best to use to determine exposure and development.

In picking which areas of a scene to previsualize as the most important shadows and highlights, keep these suggestions in mind. First, the accuracy of placement and development determination is enhanced by choosing two zones that are as far apart on the zone system scale as possible. Second, with the first suggestion in mind, certain zones are easier to previsualize than others. Zone III is an example. In every scene there should be a shadow area that you want to be the darkest in the print and still have detail. This makes it likely that one of the first areas you previsualize will be a zone III. The highlight zones that are easiest to previsualize are zone VI, which can be related to Caucasian skin, and zone VII, which is the lightest zone with texture. This makes it as easy to pick

Previsualization: *The first
step in using the zone system is
to look at the scene in front of
you and form a mental image of
the print that you wish to make
from it. This is done by using
the zone system to describe
the print. Each zone will define
a certain area, i.e., textured
highlight, detailed shadow, and
so forth. In this way the mental
image can be turned into the
technical operations such as
exposure placement and development time that are
needed to produce the final
print. Sketching the scene in a
notebook as you previsualize it
will help you relate the vision to
the craft.*

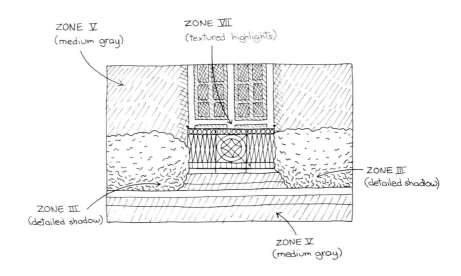

out as a zone III. In general, these zones will be easier to previsualize than the textureless zones (II and VIII) and will provide tones that are far enough apart to make the exposure placement and contrast readings accurate. It is almost never possible to get an accurate meter reading from zones I and IX since they are so extreme and usually appear in only small areas of an image.

Occasionally, it will be difficult to look at a scene and decide what areas are the most important shadows and highlights. This is usually a problem with flat scenes (N−1 or N−2). Very often, there are no distinct shadows or highlights which stand out. In this case, try squinting your eyes so that the image before you is thrown out of focus. When you are not distracted by "detail," the highlights and shadows seem to pop out and become more noticeable. Or, compare indicated meter readings of all the areas that you suspect might be the shadows and highlights. This way you can tell which area reflects the most light and judge accordingly how you want to previsualize the scene.

DETERMINING EXPOSURE
After you have previsualized a scene and established a mental image of the final print, the next step is to meter the important highlights and shadows and make a note of the indicated meter readings. (It helps to carry a pad and pencil with you when you are

starting out.) If you feel confident about your previsualization, meter only one shadow and one highlight area. If you have any doubt, then it is important to meter all the possible shadows and highlights. When comparing the reflectance values of each area it may mean that you will have to change your previsualization. A shadow that you originally saw as a zone III may reflect more light than another area that is equally important to render with detail. This would mean a change of exposure and probably also contrast. When picking a shadow value for placement be sure to consider the other shadow values in the scene and their actual relation (in terms of indicated meter readings) to each other. Your placement will decide which areas are seen as detailed shadows (zone III), open shadows (zone IV), or simple dark areas (zone II). No change in development will undo your placement of the shadow zone. Until you are comfortable with the procedure it helps to make a sketch of the scene and label the areas that are previsualized along with their indicated meter readings. This will aid your previsualization and will help you to pick the best shadow value for placement.

DETERMINING DEVELOPMENT
Since contrast is also decided by how the scene is previsualized, it is as important to accurately choose highlights for development as it is to decide the most important

Metering: *Once a scene has been previsualized the previsualized areas are metered and the indicated readings recorded. From this information the contrast of the scene is determined by comparing the number of f-stops that separate the most important highlights and the most important shadows to the number of zones previsualized between them. Exposure is found by placing the shadow reading in the zone for which it was previsualized.*

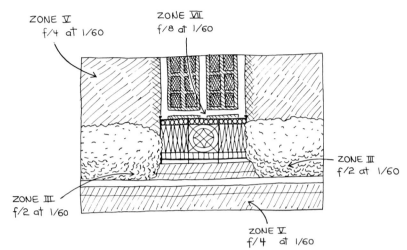

ZONE Ⅴ
f/4 at 1/60

ZONE Ⅶ
f/8 at 1/60

ZONE Ⅲ
f/2 at 1/60

ZONE Ⅲ
f/2 at 1/60

ZONE Ⅴ
f/4 at 1/60

CONTRAST is N
DEVELOPMENT is N
EXPOSURE is f/4 at 1/60 (f/2 at 1/60 placed in ZONE Ⅲ = f/4 at 1/60)

(a) Shadow Left in Zone V

(b) Shadow Placed in Zone III

Placement: *In this detail of the shadow area, print (a) is exposed for the indicated meter reading taken from that area. The result is that the shadow is rendered in zone V. Print (b) places the shadow where it was previsualized, in zone III, by stopping down the indicated meter reading two stops, one for each zone below zone V.*

shadow for exposure placement. In doing this, check several of the most important highlights with your light meter, keeping notes of the indicated readings. Before deciding upon the final development time (N, N−1, N+2, etc.), mentally visualize what will happen to the other highlights when you develop for the one you have chosen. Remember that once you have chosen a development time for a certain highlight, then the others will be lighter or darker by about one zone for each stop of reflectance difference. You may wish to make a change, either in previsualization or in the zone you develop for, or both, if what you are getting isn't what you want.

Personalizing the Zone System

The basic theories and working procedures of the zone system have now all been presented. Understanding how to relate exposure and development to the photograph that you want to make applies to all cameras in all situations. Even though it is possible to make the zone system more complicated and to adapt it to very specialized types of equipment, the fundamental ability to previsualize a scene and then translate that previsualization into actual working procedure is the real heart of using the system.

The remainder of this book outlines ways that you can personalize the zone system to your equipment, working methods, and the kinds of films that you use. The following chapters include: testing methods for determining exposure and development for black-and-white film; adapting zone system techniques to the needs of 35mm cameras with built-in light meters; and, finally, using the zone system with color films, how they differ from black-and-white and how color can change the way that you previsualize.

*The result of all of this work is
an image in which the original
vision is carried through
exactly to the final print.*

6
The First Calibration Test: Exposure Index

Once you understand the zone system as a method for seeing and making photographs, then a few tests are necessary to calibrate equipment and working procedures to the system. The two most important areas for testing are exposure and development. Since correct development depends on obtaining correct exposure, the first test will be to establish an exposure index, or "E.I."

Why an Exposure Index Test Is Necessary

Exposure index refers to the calibration of the camera to a standard film density, another way of saying correct exposure. Exposure accuracy depends on several factors. First and foremost is the ability to place a previsualized shadow zone. But correct placement is impossible if the camera and light meter are not working the way they should. This can happen for a number of reasons. Cameras and light meters rarely leave the factory perfectly calibrated. A manufacturer has what is called a "tolerance for error," which means that products do not have to be exactly right in order to be sold. Even when manufacturing tolerances are very close to the standard, subsequent shipment and handling make it necessary for you to recalibrate. Without testing, you cannot be sure that the exposures you calculate are giving you the film densities you previsualize.

For example, the standard calibration for light meters is a tone of 18% reflectance (zone V), but it is not uncommon for the actual reading to be as much as one stop away from zone V. This would make a zone III placement a zone IV (too much density and extra grain in the image) or, even worse, a zone II that has no detail. Cost does not even seem to be a factor in how accurate light meters are because too many things beyond the manufacturer's control can affect calibration.

Two possible causes for meter problems are the battery and careless use. If the meter is a battery resistance type, then the age and voltage of the battery can cause incorrect readings. The cure is prevention. Check the battery regularly, and replace it at least once a year. If you are unsure about the condition of your battery now, replace it before you begin this test. Another problem is mishandling. Pointing the camera at the sun while the meter is turned on is a common way to get underexposures. Most meter cells have a memory and will be "blinded" by the light of the sun for up to 15 minutes. The memory also affects readings when the camera is stored in the dark for long periods of time with the meter turned on. For several minutes after you bring it into the light, the readings will tend toward overexposure. Light meters in some 35mm cameras use selenium photo cells, which are less prone to the "memory" problem.

Previous page: Having the correct exposure index means that placed tones will be reproduced as previsualized. The dark bushes that form the central part of this image are an example. Had they been a zone lighter or darker than the zone III in which they were placed, the sense of detail and darkness would have been lost.

Shutters can also cause calibration problems. Coming off the assembly line, no two shutter mechanisms are exactly alike and, as they are used, gears gradually wear, springs lose their tension, and lubricant dries out and gets redistributed. Even the newer electronic shutters can give exposure times as much as 50% different from their set times. If manufacturing error is combined with wear, then the shutter loses even more accuracy. One way to slow down wear is to store your camera with the shutter uncocked. This relaxes the tension on the springs that drive the mechanism. Long periods of tension cause the springs to weaken and the timing to become erratic.

When a shutter error is compounded by a light meter error, then it is possible for a camera to be more than a stop off in exposure. This means that placement can be off by more than a zone. More than a simple aggravation, this problem can cause the zone system not to work. Every camera has a different combination of errors; no two are exactly the same. Although in some cameras the errors will actually cancel each other out, or in others the calibrations are close enough to the standards that the correct exposure can be achieved without testing, this can never be assumed. Even when the light meter and shutter are adjusted by a repair person, it is impossible to be certain that the calibrations fit your working methods.

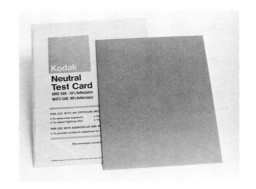

What an Exposure Index Is

One fortunate thing about the inevitability of errors is their reliability. Unless the problem is serious, most camera errors are consistent. A light meter will read consistently above or below the zone V standard, and a shutter will be consistently slow or fast.

Because of this consistency, it is possible to compensate by changing the setting of the light meter, even if the error is in the shutter. All errors are linked to the final density of the negative, and the meter is the easiest to recalibrate. This is done by making a change on the ASA dial. (ASA stands for American Standards Association; European ratings are called "DIN.") Normally the ASA dial refers only to the light sensitivity of the film you are using. The numbers double or halve as the film's light sensitivity increases or decreases by a stop. However, when the ASA setting is used to compensate for a consistent error in exposure

The 18% reflectance value of the Kodak Neutral Test Card (gray card) is the standard to which all light meters are calibrated. In fact, however, many meters vary from this standard, either because of wear or manufacturing tolerances for error.

caused by the light meter or shutter, it is called an exposure index. So, while Kodak's Tri-X film may have an ASA rating of 400, it may be necessary to set your camera's light meter to a higher or lower number in order to get the correct exposure. Calling this number an exposure index instead of ASA is not just a semantic difference. The term E.I. makes you aware that more than just a film's sensitivity to light is being considered. Instead, it is the interdependence of the entire exposing system: light meter, shutter, as well as film, that is important in obtaining the best exposure.

Testing for an Exposure Index

Many accurate testing devices are available to measure light meter settings, shutter speeds, and film densities. But most are too expensive to be practical or necessary for the 35mm photographer. The exposure index calibration test described in this chapter will produce the right exposure for placement in an empirical way because it duplicates as much as possible how you use your camera in taking pictures. This is what is meant by empirical; the test relies on practical experience rather than laboratory experiments. No special equipment is needed, only your camera and a standard black-and-white darkroom. If you have more than one camera body, each one must be tested separately;

however, different lenses usually do not, as most lens aperture settings are accurate.

MATERIALS
The following materials are needed for an exposure index test: a roll of the film that you plan to test (36 exposures), a Kodak 18% gray card, an object with zone III detail like a dark blue or black sweater, something with zone VII texture like a white sweater or a white terrycloth towel, and finally a willing and patient subject. A tripod is useful, but optional.

PREPARATION
Running the E.I. test is simple. First, take the subject outside into normal contrast light, such as the north side of a house on a sunny day. However, it could be anywhere that the light meter indicates a four-stop difference between the zone III and zone VII values that you have brought along. Even though the light meter isn't calibrated yet, the contrast readings will be accurate because any error will be consistent. Stand the subject against a neutral background wearing or draped with the zone III and zone VII tones and holding the gray card. Compose a "head and shoulder" portrait in your viewfinder, filling the frame with as much of your subject's face as possible and still showing the zone III, zone VII, and the zone V gray card in the frame. This composition should be used for the whole test. You may want to

For the exposure index test, compose a "head-and-shoulders" portrait that includes the previsualized highlight, shadow, the gray card, and plenty of your subject's face. After metering the most important highlight and shadow values to determine a normal contrast, proceed to place the shadow reading for each index to be tested. In this image, the hair is previsualized as zone III, the sweater as zone IV, the face as zone VI, and the blouse as zone VII. The indicated meter reading of the hair was used to place the exposure.

use a tripod, but it is not necessary unless the exposure times are longer than 1/60 of a second.

With the camera loaded, set the ASA dial on the light meter to the film's ASA rating. Since all the exposures made on this roll will be tests to find the correct meter setting for your exposure index, the numbers on the ASA dial will now be referred to as "E.I.'s." Next, meter the zone III tone following the procedure and observing the cautions outlined in Chapter 3, and place the indicated meter reading in zone III. Keep careful notes of all your indicated meter readings and exposure placements throughout the test. Should something go wrong and show up later in the test, your notes will prove invaluable in locating the problem. Never trust your memory.

At this point you are ready to begin shooting your test.

Procedure

FRAME 1: Shoot at the same E.I. as the film's ASA.

FRAME 2: Cover the lens and shoot a blank. This will help avoid confusion later on, although it is not absolutely necessary.

FRAME 3: Set the E.I. to 4 times the film's ASA. Then re-meter the zone III tone and place it exactly as you did for the first frame. Make a note of the indicated meter reading and the exposure placement. If the light hasn't changed from the time you metered the first frame, then the

On the next two pages are sample records that outline the procedure for completing an exposure index test. One is filled out with the data recorded from the test used as the illustration in this chapter; the other is left blank and can be photocopied for your personal use. These forms contain all the information needed to evaluate the negatives and organize your results clearly. An added advantage of recording all of this information as you shoot is that it forces you to work more slowly and to catch careless errors.

EXPOSURE RECORD: *E.I Test*

DATE: *10-3-82*   FILM USED: *Kodak Tri-X (ASA 400)*

CAMERA USED: *Nikon - ser. # 7253119*

COMMENTS: *Previsualized hair as Zone III, sweater as Zone IV, skin tone (face) as Zone VI,*
blouse as Zone VII, Indicated meter readings (w/meter set at film's ASA) were f/4 at 1/125 sec.
for hair and f/16 at 1/125 sec. for blouse. — Normal Contrast Scene —

Frame	Procedure	Indicated meter reading off of zone III shadow value	Calculated exposure (placed in zone III)
#1	Shoot at the same E.I. as the film's ASA.	*f/4 at 1/125 sec.*	*f/8 at 1/125 sec.*
#2	Cover the lens and shoot a blank.		
#3	Set the E.I. to 4 times the film's ASA.	*f/8 at 1/125 sec.*	*f/16 at 1/125 sec.*
#4	Shoot a blank.		
#5	E.I. at 2 times the film's ASA.	*f/5.6 at 1/125 sec.*	*f/11 at 1/125 sec.*
#6	Shoot a blank.		
#7	E.I. at the same rating as the film's ASA.	*f/4 at 1/125 sec.*	*f/8 at 1/125 sec.*
#8	Shoot a blank.		
#9	E.I. at 1/2 the film's ASA.	*f/2.8 at 1/125 sec.*	*f/5.6 at 1/125 sec.*
#10	Shoot a blank.		
#11	E.I. at 1/4 the film's ASA.	*f/2.8 at 1/60 sec.*	*f/5.6 at 1/60 sec.*
#12	Shoot a blank.		
#13	E.I. at 1/8 the film's ASA.	*f/2.8 at 1/30 sec.*	*f/5.6 at 1/30 sec.**
#14	Shoot a blank.		
#15	E.I. at 1/16 the film's ASA.	*f/2.8 at 1/15 sec.*	*f/5.6 at 1/15 sec.**
#16	Shoot a blank.		

After shooting: Review your notes. Reshoot
any exposures that don't look correct. It is
easier to do it now than start over again.

* Note: Used tripod for these frames.

#17	Reshoot E.I. _____.
#18	Shoot a blank.
#19	Reshoot E.I. _____.
#20	Shoot a blank.
#21	Reshoot E.I. _____.
#22	Shoot a blank.

Exposures look OK.

EXPOSURE RECORD: _____

DATE: _____ FILM USED: _____

CAMERA USED: _____

COMMENTS: _____

_____

_____

Frame	Procedure	Indicated meter reading off of zone III shadow value	Calculated exposure (placed in zone III)
#1	Shoot at the same E.I. as the film's ASA.		
#2	Cover the lens and shoot a blank.		
#3	Set the E.I. to 4 times the film's ASA.		
#4	Shoot a blank.		
#5	E.I. at 2 times the film's ASA.		
#6	Shoot a blank.		
#7	E.I. at the same rating as the film's ASA.		
#8	Shoot a blank.		
#9	E.I. at 1/2 the film's ASA.		
#10	Shoot a blank.		
#11	E.I. at 1/4 the film's ASA.		
#12	Shoot a blank.		
#13	E.I. at 1/8 the film's ASA.		
#14	Shoot a blank.		
#15	E.I. at 1/16 the film's ASA.		
#16	Shoot a blank.		
	After shooting: Review your notes. Reshoot any exposures that don't look correct. It is easier to do it now than start over again.		
#17	Reshoot E.I. _____.		
#18	Shoot a blank.		
#19	Reshoot E.I. _____.		
#20	Shoot a blank.		
#21	Reshoot E.I. _____.		
#22	Shoot a blank.		

exposure should be two stops less. If it isn't, then it indicates either a procedural error or an inconsistency in the meter that needs repairing. Check your procedures first.

FRAME 4: Shoot a blank.

FRAME 5: E.I. at 2 times the film's ASA. Re-meter and shoot as in frame 3.

FRAME 6: Shoot a blank.

FRAME 7: E.I. at the same rating as the film's ASA. This should be a duplicate of frame 1. This is another opportunity to check procedure and the consistency of the light meter.

FRAME 8: Shoot a blank.

FRAME 9: E.I. at 1/2 the film's ASA.

FRAME 10: Shoot a blank.

FRAME 11: E.I. at 1/4 the film's ASA.

FRAME 12: Shoot a blank.

FRAME 13: E.I. at 1/8 the film's ASA.

FRAME 14: Shoot a blank.

FRAME 15: E.I. at 1/16 the film's ASA.

FRAME 16: Shoot a blank.

After shooting this series, study your notes. If you had been testing a fast black-and-white film (ASA 400), then the E.I.'s you should have been using would be 400 (frames 1 and 7), 1600 (frame 3), 800 (frame 5), 200 (frame 9), 100 (frame 11), 50 (frame 13), and 25 (frame 15). Films having other ASA ratings would have a similar series of E.I. settings based on their emulsion speeds. Your notes should also indicate a relationship between each exposure of one stop, except between frames 1 and 3, which should have been two stops apart. Only a departure from this relationship caused by changing light, such as on a partly cloudy day, is acceptable. Otherwise use the remainder of the roll to shoot the test over. If the information in your notes looks correct, then reset your camera's meter to an E.I. that is the same as the film's ASA and begin to shoot the rest of the roll.

Expose the film in this manner. First choose a shadow value in the scene you want to photograph. Previsualize it (zone II, III, or IV), place it in the previsualized zone, and shoot. Then bracket that image by exposing it one stop more and one stop less than the placed exposure. This is the equivalent of exposing the same scene at three different E.I.'s. Finish the rest of the roll shooting scenes that interest you. Contrast does not matter, but it would be good practice to meter both highlights and shadows keeping notes about contrast and exposure for all scenes.

FILM PROCESSING
After you expose the film, process it as you normally would. Use the manufacturer's recommendations for time and temperature if you do not have any other information. Agitation of the film is important and

*When the film from the expo-
sure index test is developed,
examine it carefully and identify the index being tested for
each frame. What you see in the
negatives at this point will be
the first indication of which
negative is correctly exposed.*

should be consistent no matter
what method you use. Complete
tank inversions are recommended
at the rate of six to twelve every 10
seconds. Normally, agitation is constant for the first minute and then
for 10 seconds every minute after
the first. Also, temperature is important. Be sure you have an accurate thermometer. Dial-type ther-
mometers are never accurate. If you
use one, check it against a glass
mercury type periodically before
you rely on the readings.

INSPECTING THE NEGATIVES
After the processed film dries, carefully examine the negative densities. A magnifier and an even
source of light will help. Take the
magnifier and look for detail in the
zone III area of frame 1. If there is

detail, then that area of the negative
fits the definition of zone III. Next
look at frame 3. This frame, exposed
two stops less than frame 1, should
be lacking detail in the same zone
III area. Continue looking at the
other frames in the order that they
were shot until you find the first
frame that has detail in the zone III
area of the negative. This is a preliminary indication of the correct
index as the first frame to show the
proper zone III density is usually
the one shot at the correct E.I. It is
not final, however, as there can be
visible density in the negative that
will not show up in the print. If no
zone III detail appears in frame 1,
then expect detail not to show up
until at least frame 9, and the E.I. to
be less than the film's ASA. If detail
appears in frame 3, it indicates that

*Begin your examination of the
negatives by arranging them in
order from the least exposure
(highest E.I.) to the greatest exposure (lowest E.I.). Look at the
previsualized shadow zones.
As the index is lowered, exposure increases and the zone III
area begins to show detail. The
negative that first shows
adequate detail in the zone III
area is a preliminary indication of the correct index.
In this series of negatives,
frames 3 (E.I. 1600) and 5 (E.I.
800) do not show any detail in
the hair (the previsualized and
placed zone III value). Frames
7 (E.I. 400) and 9 (E.I. 200) show
adequate detail and indicate
that the correct index is probably in the range of 400 to 200.
Frames 11 (E.I. 100), 13 (E.I. 50),
and 15 (E.I. 25) show detail in
the zone III area, but also have
extra density which will show
up as larger grain on the print.*

Frame 3, E.I. 1600

Frame 11, E.I. 100

Frame 5, E.I. 800

Frame 13, E.I. 50

Frame 7, E.I. 400

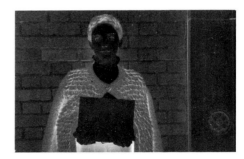

Frame 15, E.I. 25

Frame 9, E.I. 200

the E.I. will be much higher than the ASA. In this case, a new test must be shot at E.I.'s 8 times and 16 times the film's rated ASA to be sure of accuracy.

PROPER PROOF

The next step is to make a contact sheet of the negatives. This is done by using the zone system. We know that a correctly exposed zone I should print as a maximum black, the same as the clear film base. If we expose our contact sheet with enough light to print the film base as maximum black, then correctly exposed tones on the negative will show up in the proper zones on the print. A simple way to be sure is to look for the sprocket holes on the film to disappear into the same print tone as the film base. (See illustration on p. 68.) In addition to the correctly exposed frames looking right, any underexposed frames will appear too dark and any overexposed frames will appear too light. In this way, the contact sheet should confirm that the frame selected in the negative inspection has the best appearance. Making a contact sheet any other way does not give you this information. This is why it is sometimes called a "Proper Proof." If you find a wide variation between the results of your negative inspection and the proper proof, review all your procedures.

PRINTING THE E.I. TEST

The best way to evaluate the negatives from the E.I. test is by enlarging them. Starting with frame 1 (E.I. same as the film's ASA), make standard size enlargements in which you try as carefully as possible to match the skin tones of the print to your subject's. This will not be difficult for Caucasian skin as it all fits into zone VI, and, as noted, most of us tend to be very aware of how skin tones (especially the face) are rendered in black and white. If the tone is printed a little dark or a little light we usually notice it.

After making the first print, all the other negatives in the E.I. test are printed, adjusting the exposure so that in each the tones of the face match. You may skip frame 7 as this should be a duplicate of frame 1. Be sure to make all prints on normal contrast paper and always print the full negative without cropping. When the prints are dry it is time for the final determination of your exposure index. One of these prints will look better than the others.

EVALUATION OF THE E.I. TEST

Spread the prints out in a well-lit space where you can see all of them at once. Avoid daylight only or fluorescent only lighting situations. Put the prints in order from the highest E.I. (4 times the film's ASA) on the left to the lowest E.I. (1/16 the film's ASA) on the right. Then, check to see that the faces in all the prints match the actual skin tone and also each other. If they are slightly off this standard, it is possible to make mental allowances when examining the prints. If, however, they are more than a little off, either with re-

*a. A correctly made contact
sheet can give information
about exposure as well as
showing what the negatives
look like. The "proper proof" is
made by exposing the film base
for a zone I value. The sprocket
holes will literally disappear as
the exposure reaches the point
that the film base prints as
maximum black. In this example 25 seconds is the correct
exposure time. Before that, the
sprocket holes are clearly
visible, after that is simply an
indication of excessive
exposure.*

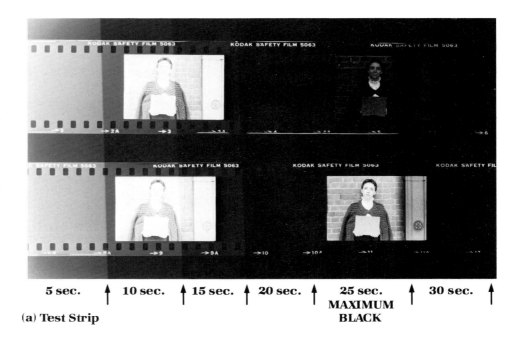

5 sec. ↑ 10 sec. ↑ 15 sec. ↑ 20 sec. ↑ 25 sec. ↑ 30 sec. ↑
MAXIMUM
(a) Test Strip BLACK

spect to each other or to the actual skin tone, then it will be better to go back and reprint the ones that are not correct. A little extra time spent here will prevent having to rely on guesswork later on.

Begin your examination of the prints with the zone III area. Starting at the left, eliminate from further consideration any prints that do not show enough detail in the shadows. You should already have an expectation of which prints these will be based on your examination of the negatives, but try to be objective. However, if you find yourself keeping prints made from negatives that showed no detail or eliminating prints made from negatives that showed definite detail, then stop and review, and do not proceed until you can discover a

reason for this. If you do find a mistake, do not continue until you have corrected it.

Next, examine the highlights. This time, starting with the prints on the right (lowest E.I.'s) eliminate any prints that are too dark or gray in the zone VII area. This should be easy when you compare them with all the remaining prints. Keep only those prints that you feel have an acceptable highlight.

By this time you should have narrowed the original seven prints down to only two or three. If not, review the first two steps. The final determination will be made by carefully inspecting the grain of the prints as well as the overall image quality. In looking for the best grain structure, look closely at the face and the gray card. The least mottled

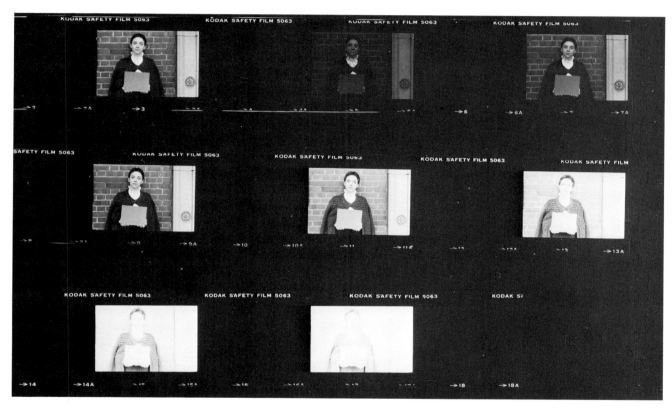

(b) "Proper Proof"

skin and the smoothest gray card will indicate the finest grain. If these areas do not help, look at any relatively smooth area in zones IV-VI.

If, after eliminating any prints that show obviously inferior grain structure, more than one print is left, then look at the overall picture. Keep in mind the qualities of the various tones you have purposely included in the image. Try to see if one print shows all of these tones in the best possible way. Don't be misled by a good rendering of just one zone—find the print that is best in all areas combined. When all but

one print is eliminated, the E.I. at which the image was made is your exposure index. If, as sometimes happens, you cannot decide between two adjacent E.I. prints, then you may assume that your true E.I. is halfway in between these two.

The exposure index that you find in this test may be the same as the manufacturer's ASA rating for the film (experience shows that approximately half the E.I.'s are), or it may be different. Whatever the case, it is now personalized to your equipment, and allows you to know that when you place a shadow it will ac-

b. In the proper proof, the underexposed frames (numbers 3 and 5) appear too dark. The overexposed frames (numbers 11, 13, and 15) are too light. Frames 1, 7, and 9, seem to be the best. These are also the frames which had the correct amount of shadow detail in the negatives. This is another indication of the correct index, somewhere in the 400 to 200 range.

a. E.I. 1600:
*Shadow values in the hair
(zone III) are too gray and
lacking in detail.*

b. E.I. 800:
*While the hair is darker in
tone than in the E.I. 1600
print, it still lacks detail.*

c. E.I. 400:
*Satisfactory shadow detail
(in the hair); Satisfactory
highlight tone (in the zone
VII blouse); Least amount of
grain in the face and gray
card; Best overall appearance of tones.*

d. E.I. 200:
*Satisfactory shadow detail;
Satisfactory highlight tone;
Some appearance of grain in
the face and gray card; Not
quite as much contrast as in
the E.I. 400 print.*

e. E.I. 100:
*Satisfactory shadow detail;
Satisfactory highlight tone;
Noticeable grain in skin and
gray card.*

f. E.I. 50:
*Satisfactory shadow detail;
Highlight of zone VII blouse
is too gray.*

g. E.I. 25:
*Satisfactory shadow detail;
Highlight of zone VII blouse
is too gray.*

*The final step of the exposure
index test depends on the
printing of the skin tones of the
subject. Be certain that the
faces in the prints match as
closely as possible before
you begin your evaluation.
In this example, the final
choice was between the E.I. 200
and the E.I. 400 prints as the
best overall. It was the personal
choice of the photographer to
pick the E.I. 400 print as having
the most pleasing tones overall. Another choice that was
possible if neither print had
been significantly better than
the other would have been to
choose an E.I. of 300.*

*Halftone reproduction makes
it impossible to show all of the
detail present in the original
prints. This might make it difficult for you to see the subtle
differences mentioned in the
text.*

tually be in the zone you pre-visualize. This test works because film's response to exposure is exactly as the zone system predicts. An underexposed zone III will lack shadow detail in both the negative and the print. In the same manner, an overexposed negative will have highlights that are pushed against the negative's maximum density. Highlights printed from an overexposed negative will appear flat and darker than previsualized.

Grain is also a product of exposure. The more density that a negative has, the more grains of silver that are clumped together, and the more you will notice that clumping in the print. The optimum negative desired in this test is one that has enough density to show detail in placed zone III areas, but not so much density as to cause excess grain or gray highlights. Even though a printable negative can be gotten with a wide variety of exposures as shown in this test, only one exposure can produce the best possible print when it is closely examined.

A densitometer can provide the information needed to calculate an exposure index directly from the film. However, density information from this machine cannot indicate what an image will actually look like.

Alternatives to the Standard E.I. Test

The best tests in photography are empirical tests which duplicate actual working conditions. The E.I. test described in Chapter 6 is such a test. However, other methods of finding an exposure index are worth mentioning.

1. If you have access to a densitometer (a device that measures film density—many custom labs and university photography departments have them), then readings made right on the film will give you information needed to calculate an exposure index. For this method, simply alternate a blank frame with a frame shot of a gray card (out of focus and filling the viewfinder) placed on zone I. Each frame of the zone I tests a different E.I., just as in the test in this chapter. When the film is developed and dry, take densitometer readings from the film base (blank frames) and the zone I values. The correct exposure index is indicated by the zone I frame which reads a density that is between .03 and .06 higher than the film base reading. If you are testing 120 roll film or sheet film, the density difference should be .10 to .15 higher than the film base.

2. If you don't have access to a densitometer, an alternative to the above test is to shoot a roll of film in a similar manner (using the same E.I.'s) as the above test, only placing

Film Base **Zone II Exposure**

25 sec.

20 sec.

15 sec.

10 sec.

5 sec.

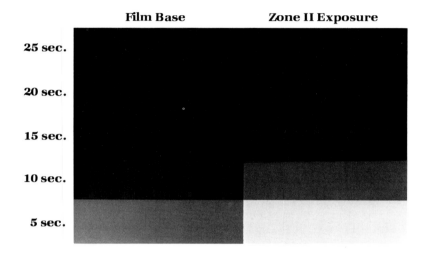

the gray card in zone II instead of zone I. Examine the processed film and select the first frame that is visible (some of the underexposed frames will have no visible density). Place it in the negative carrier of your enlarger so that half of the adjacent blank frame is in the opening along with the zone II frame. With the enlarger focused to make an approximate 8 x 10″ print, and the lens stopped down to around f/8, expose a strip of #2 enlarging paper at increasing 5-second intervals, covering a small section at each interval. The developed strip will show the effect of increasing exposures on both the film base and the zone II density that is being tested.

When the print is dry, examine the effect of increasing exposure on the film base side. Look for the step at which the film base prints as maximum black. At this point the enlarging paper will not get any darker even though exposure may increase. If this doesn't happen, even at the longest exposure time, make another strip, either increasing the time or opening the enlarging lens one stop. At the point where the film base density prints as maximum black, look at the zone II side of the strip for the same exposure time. It should show a definite lighter tone, just as zone II is a definite lighter tone than maximum black. If it does show this difference at a normal viewing distance and under normal room lights, then this frame represents the correct E.I. If it does not, it indicates that even though visible, the density isn't enough to print as a separate tone the way a zone II should. This test assumes that you are using fresh developer at the correct temperature, and fresh enlarging paper to make all prints.

7

The Second Calibration Test: Development Time

Developing for the Highlights

If exposure is the first part of the zone system equation, then development is the second. Once the correct exposure calibration is found in the E.I. test, the next step is to determine development times for normal, N+, and N− contrast. As we have seen in the graphs in Chapter 5, the important consideration in film development is to stop it at the point that the contrast range of the tones on the negative match our previsualization.

Since, in theory, the development of every negative varies according to the previsualization, it is necessary to test for each type of contrast that might be encountered. No mathematical formula can take into account the differences in films and developers, not to mention the variations in individual processing methods. Fortunately, the testing is simple once you apply the zone system to the problem. It is a matter of defining the needs of the scene in terms of zones and then using the relationship of contrast to development time to find those specific tones in the print.

More simply put, once you have previsualized a scene, you know how you want the highlights and shadows to look on the print. Then, exposure placement determines the shadow densities on the negative. They will not change significantly during development. Finally, development times need to be

tested until one is found that renders the important highlight density the way it was previsualized. The test is simple, but finding the correct highlight density this way is time-consuming. Plan to set aside a few hours each day for the next several days to complete the development test.

Finding Normal Development Time

The first part of this test is to find a normal contrast subject. Look around with your previsualization and light meter until you find an appropriate scene. It should have large, easily recognizable areas of shadow, especially zone III, and an equally large area of highlights, especially zones VII and VIII. Place the exposure carefully for the most important shadow value, and determine a normal contrast range between the placed shadow zone and the most important highlight zone. Then expose an entire roll of film. A tripod will help make each frame as nearly identical as possible.

In the darkroom cut the film into approximately three equal lengths. Save two in a light-tight container, and develop the third for the manufacturer's recommended development time, or whatever you consider to be a normal development time. When the film is dry, test the contrast by printing one of the frames on normal contrast paper. Using a tray of fresh developer, ex-

Previous page: In this image the grass was previsualized and placed in zone III rather than in zone V, which is the normal rendering of sunlit grass. The benches were given their luminous quality by longer-than-normal development. Once the exposure and development tests are completed, such nonstandard previsualizations will present no more difficulties than conventional ways of seeing the world.

SHIRT
ZONE VII previsualization
Indicated meter reading
f/22 at 1/125

GRASS
ZONE V previsualization
Indicated meter reading
f/11 at 1/125

PANTS
ZONE III previsualization
Indicated meter reading
f/5.6 at 1/125

NOTE: 4 stop range between zones III and VII indicates
normal development. Exposure calculated to be f/11 at
1/125 (placement of pants in zone III).

Every negative is developed according to its previsualization. When the development time is correct, the tones will appear as they are seen by the photographer. In this sequence the previsualization was for a normal development. The negatives from this scene were developed for N, N−1, and N+1. When printed for correct zone VII highlights, the change in contrast created by development is shown by changes in the shadow densities. The previsualized zone III shadow is too gray in the N−1 negative and too dark and lacking in detail in the N+1 negative. Only the normally developed negative prints the tones in the way that they were previsualized.

N−1 Development

N Development

N+1 Development

pose and develop the print so that the important highlight is properly rendered. Then, once the print is dry, examine the placed shadow to see if it fits the correct description for that zone. For example, does a zone III have detail or not? If it is too dark, that indicates that the negative is overdeveloped and too contrasty. Or, perhaps a zone II appears gray and not dark enough, an indication of underdevelopment and too little contrast in the negative. If both the highlights and the shadows appear the way you previsualized them, then the development time is correct for normal contrast light.

Reducing Development

If you determined that the first segment of film was developed too long, then take the second segment and develop it for less time. How much less will depend on your perception of how much extra contrast appeared in the print. If the zone IV areas of the print looked like a zone III and the zone III areas had no detail at all, then try developing about 7% less for low-speed film (ASA 25–50), 10% less for medium-speed film (ASA 100–200), and 15–20% less for high-speed film (ASA 400). If the zone IV areas of the print were too dark even to have detail, then try doubling the percentages. If the print was just a little too contrasty, such as the zone III having a slight but not really sufficient amount of detail, then try reducing development by less than the above percentages. A rule of thumb is that the lower the light sensitivity of a film emulsion the more rapidly it reacts to changes in development.

After testing the second segment of film, you have the third segment available if you need to make further adjustments in development time, or as a backup if something goes wrong in processing one of the other two.

Increasing Development

If you determined that the first segment of film was underdeveloped, and the print too flat, then increase the development of the second segment by the same percentages that you would have decreased it for too much contrast. Again, look at the way the shadow values are rendered in the print. If they are too light for the zone that they were previsualized in, the development time needs to be longer. By being careful and controlling temperature and agitation, you should be very close to finding the correct development time by the third segment, if not the second.

It might seem confusing that the development time of the film is evaluated by looking at the shadows of the print, especially since it has been stressed that development time affects the highlights. However, there is a logical reason. What we are looking for when we make a print of the test negative is the overall contrast. This is best evaluated when the exposure of the print is correct for the highlights. The highlight of the print is the area of least density and, like the shadows in the negative, is the area most sensitive to changes in exposure. Therefore, it is the gauge of correct exposure. Only in a correctly exposed print can we judge the overall effect of development on contrast. Even though there may be detail in the zone III area of an overdeveloped negative, it will be suppressed in a correctly exposed print made on normal contrast paper. You will notice a similar effect in an underdeveloped negative, where the shadows will print lighter than they are pre-visualized.

Finding N+ and N− Development Times

The tests for finding development times to fit N+ and N− contrast are similar to the test for normal contrast. Find a scene with the type of contrast you wish to test. As in the normal contrast scene, make sure that it has large areas of easily iden-

tifiable zones II and III shadows and zones VII and VIII highlights. This will make the evaluation of the prints easier. Next shoot an entire roll of the scene, being sure to make the most accurate exposure placement possible. For N + 1 contrast, begin with a development time that is longer than the normal time you found by approximately 20% for high-speed film, 10% for medium-speed film, or 7% for slow-speed film. For N − 1 contrast use the same percentages to reduce the de-

ZONE III
Indicated meter reading
f/5.6 at 1/125

ZONE VII
Indicated meter reading
f/16 at 1/125

NOTE: 5 zones difference in previsualization and 4 stops difference in indicated meter readings. Exposure is f/11 at 1/125 (placing zone III value), contrast and development is N+1.

In a low contrast scene such as this, more development (N+1) is needed to increase the separation between the highlights and shadows. At normal development time, the zone III appears on the print more like a zone IV.

N Development

N+1 Development

*For a scene that is previsual-
ized to have high contrast, a
lower-than-normal development time is needed. In this
case, reduced development
(N−1) allows the shadow detail
in the previsualized zone III to
be seen in the print.*

ZONE IV
Indicated meter reading
f/2.8 at 1/125

ZONE VIII
Indicated meter reading
f/16 at 1/250

NOTE: 4 zones difference in previsualization and
5 stops difference in indicated meter readings.
Exposure is f/4 at 1/125 (placing zone IV value),
contrast and development is N-1.

N Development

N−1 Development

velopment time from the normal. If
you are careful, these tests will be
completed by the second or third
segment of film, reducing the
amount of time spent testing. For
35mm film, it is rarely necessary to
have more than one development
time for N, N+1, and N−1 contrast.
If you ever need other development
times, they can be found later.

Empirical Testing for Contrast

It is important to realize that like
the E.I. test, this is an empirical
method. The results are geared to
how you actually make and evaluate your own prints. Therefore, the
development times that you find
are in a sense personalized to how
you like to see prints made at the
time you did the testing. Always
give yourself the freedom to change.
You might, in six months time, want
to see a zone VIII highlight slightly
lighter and more luminous on the
print, or perhaps a zone III shadow
with less detail and more density.
Remember that each zone is a segment of the continuous tone scale
and can have many different ways
to be seen in the final print. It is true
that these are subtle differences,
but as your sensitivity to the possibilities grows, your attitudes toward what you want will change. It

is likely that your development times will also have to change in response to this growth.

The Effect of Materials on Testing

In performing calibration tests for the zone system, it is important to realize that you are calibrating for the materials that you use as well as for the camera shutter and light meter. Changing film or developers can mean an entirely new set of calibration tests. Even the type of enlarging paper you use makes a difference. You should carefully select the type of film, film developer, and enlarging paper for your tests and plan to use them consistently afterwards.

In selecting black-and-white films, the basic choices are in emulsion speed. The important factors to consider are the grain size, which is finer in the slower speed films, and the latitude for controlling contrast, which is greater in higher speed films. Brand names are unimportant, except for their availability in your area.

Developers, too, have certain basic characteristics. Standard developers, such as Kodak's D-76 and Ilford's ID-11, offer fine grain and fairly flexible contrast control. Ultra-fine grain developers, such as Kodak's Microdol-X and Edwal's FG-7 with added sodium sulfite, offer the appearance of smoother grain at the sacrifice of image

sharpness (the technical term is "actuance"). High actuance developers offer the sharpest possible image, but also make grain appear more clearly defined. Examples of such developers are Beseler's Ultrafin and Agfa's Rodinal (when used at high dilutions). The combination of film and developer is important. Each combination produces an image with a different character. It takes time and consistent use of a film and developer to learn these qualities. Switching back and forth between films and developers will slow down the learning process.

Enlarging paper also has certain characteristics that affect your prints. Normal contrast grades of different brands of paper can produce slightly different renderings of tones. If you make development tests on one brand of paper, then expect some changes in the contrast if you switch to another. The differences may be subtle, but, as you will notice, one of the benefits of the testing in this book will be to increase your perception of these differences.

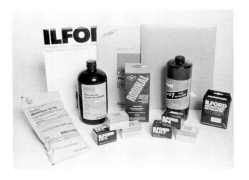

All photographic images are products of the materials used in creating them. Choice of film, developer, and enlarging paper should be made carefully and then they should be used consistently. Otherwise, the special character that a certain combination of materials gives an image will not be learned as quickly as it could.

*The type of enlarger that you
use will determine the correct
development time for your
negatives. Condenser enlargers require a lower contrast
negative than diffusion enlargers. The examples shown are
made from a negative de-
veloped for a condenser enlarger. Print (a) is made with an
Omega condenser enlarger.
Print (b) is made on the same
enlarger modified with a "cold
light" diffusion light source
manufactured by Aristo Grid
Lamp Products (Port Washington, NY). As you can see, what is
correct for the condenser enlarger is too flat for the diffusion model.*

Alternatives to the Development Test

1. With a densitometer the development test consists of taking
density readings from a zone VIII
tone. Zone VIII is the best value to
use in a densitometer test because
it is the highest negative density
that has a visible tone on enlarging
paper. Therefore, it is the tone most
affected by changes in development. Once you find the correct developing time for a zone VIII density, all other highlight zones will
also have the right density.

What constitutes a proper zone
VIII depends on the method you
use to print the negatives. Condenser enlargers (the most popular
type) render contrast differently
than the diffusion or cold light
models. The best way to find a zone
VIII density for your equipment is to
take densitometer readings on at
least ten of your negatives that have
good printing zone VIII values in
them. Once the unusually high or
low readings are thrown out, the
rest are averaged. This is the zone

VIII density that works for your system. Starting points for comparison
would be a zone VIII density of 1.15
for 35mm negatives enlarged on a
condenser enlarger and 1.25 for a
diffusion or cold light model. Zone
VIII density figures for 120 size roll
film and all sheet film would be 1.30
for condenser and 1.40 for diffusion
and cold light systems.

The testing procedure for normal
contrast involves shooting a gray
card (out of focus and filling the
viewfinder) placed in zone VIII.
Then, cutting the film up into
thirds, develop the first segment for
what is thought to be the normal
development time. If the density
reading is higher than the target
zone VIII value, develop the next
segment for less time. If the reading
is lower, develop the next segment
longer. A density difference of .30
from the target density suggests
that the next development time
should change by 7% for low-speed
film, 10% for medium-speed film,
and 15–20% for high-speed film.

Testing for N + 1 development involves the same procedure, except

(a) Condenser Enlarger

(b) Diffusion Enlarger

that the placed zone should be zone VII. Look for a development time that develops the placed zone VII to the target zone VIII density. The test for $N-1$ development requires that the gray card be placed on zone IX. The development time you are looking for is the time that will develop the zone IX exposure to the target density for zone VIII.

2. Without a densitometer, run the test in this manner. For normal contrast, alternate a blank frame with a gray card placed on zone VIII for an entire roll. After developing the first third of the roll, place it in the negative carrier of your enlarger so that half of the blank frame is in the opening along with a zone VIII frame. With the enlarger focused to make an approximate 8 x 10″ size print, and the lens stopped down to around f/8, expose a strip of normal contrast enlarging paper at increasing 5-second intervals, covering a small section at each interval.

When the strip is processed and dry, look for the exposure step at which the film base (blank frame) prints as maximum black. If none of the steps are maximum black, repeat the test at longer exposure times. At the exposure where the film base prints as maximum black, look at the zone VIII. It should begin to show a slight gray tone on the paper (which is the zone system definition of zone VIII). If the first exposure to produce maximum black for film base is also the first exposure to produce a zone VIII

gray tone, then the development time is correct. If the zone VIII gray tone appears at an exposure longer than the time required to produce a maximum black for the film base, then the development time is too long and there is too much contrast. If the zone VIII tone appears at an exposure less than the time needed to produce a maximum black for the film base, then the development time is too short. Adjust the development time for the next segments of film until the zone VIII tone appears at the same time that the film base prints as maximum black.

In the same manner, $N+1$ is found by placing the gray card in zone VII and developing it until it produces a zone VIII tone opposite the film base's maximum black. $N-1$ is found by placing the gray card on zone IX, and developing it so that a zone VIII tone is produced under the same conditions.

The Zone System for 35mm
Photographers

(a) **Underdevelopment:** *When there is underdevelopment in the optional development test, the appearance of a zone VIII tone will occur before the exposure of the film base produces a maximum black. This indicates that the next piece of film should be developed longer. How much longer is determined by the type of film you are testing and your perception of how close the differences are.*

(b) **Correct Development:** *At the correct development time the zone VIII tone that you are trying to produce will occur at the same exposure that produces a maximum black on the film base side of the test film.*

(c) **Overdevelopment:** *Overdevelopment is indicated when the zone VIII tone does not appear until after the film base has printed with a maximum black tone. When this happens, the next segment of film to be tested must be developed for less time.*

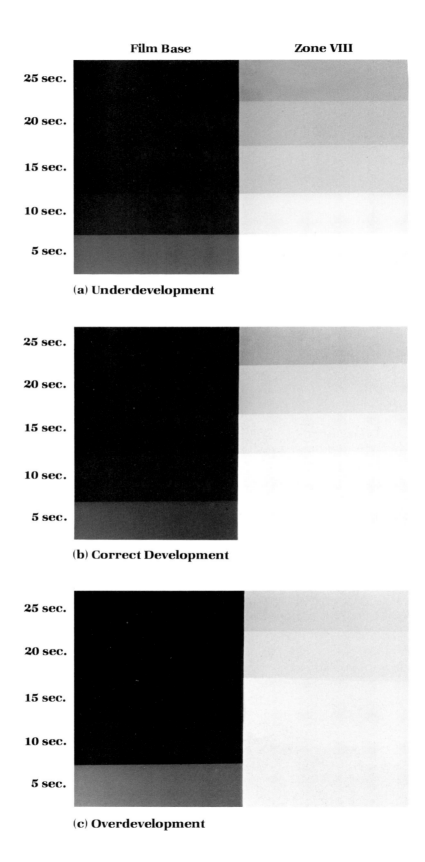

Film Base   **Zone VIII**

(a) **Underdevelopment**

(b) **Correct Development**

(c) **Overdevelopment**

8
A Working Method
for 35mm Photographers

If the zone system truly offers a grasp of the creative potential of photography, then a complete understanding of the process should reveal many ways that previsualization, metering, and contrast determination can be streamlined and made more adaptable to different ways of working. For example, to some photographers, speed is not important. Large-format cameras used on a tripod are usually photographing subjects that are not moving, or do not depend on the mobility of a light-weight camera. In these cases, you have time to previsualize all the important tones in the scene. Often, special hand-held light meters make metering these tones an easy matter. But, 35mm cameras have the special ability to move rapidly with changing scenes that comes from the lightness and portability of a self-contained unit. It is tempting, when the action is fast and furious, to give up the control of the zone system for convenience. That is not necessary. The zone system can also adapt hand cameras with built-in meters to fast-paced action.

The System

Card players, race handicappers, and stock market analysts all have short-cut systems. So does the zone system. This method involves previsualizing and metering a close, easily accessible object, one that is near you all of the time. The expo-

sure, once found, needs to be changed only when light intensity changes. Making a contrast determination occurs only at the beginning of each roll. When combined with other short-cuts, such as zone focusing or setting the lens on hyperfocal distance[1], the zone system can simplify one's technique. Rather than having to constantly make decisions about exposure and development, it allows you to concentrate on the image in the viewfinder.

Skin Tone

This short-cut works because we normally previsualize Caucasian skin in a certain way. In the brightest light available for a photograph, skin tone is seen as a zone VI. This is a standard value. Most of us have an exact idea of what skin looks like on a black-and-white print. Even people who are not photographers can sense whether or

Previous page: In some scenes it is nearly impossible to walk up and meter specific tones. If a previsualized substitute is nearby and it is in the same light as the scene being photographed, then both exposure and contrast readings will be accurate.

---

[1]Zone focusing simply means setting the focus control of the lens at an estimated distance for the main subject and allowing depth of field to make up for slight inaccuracies. This saves having to constantly change focus when the subjects are all about the same distance to the lens. Hyperfocal distance is a variation of this. Using depth-of-field scales marked on the lens barrel, it means setting the focus for the closest distance at which the depth of field also includes the infinity mark. On a bright day, with the lens stopped all the way down, the depth of field might include everything from about 10 feet to infinity.

Only a few tones in the zone system are considered to be standard values. This means that to change the way these tones are rendered in the print will change our perception of a normal scene. Caucasian skin tone is such a standard. It is always a zone VI for a normal previsualization in even light. But, when the light is uneven, as in this example, only the most brightly lit skin tone is a zone VI. The shaded skin tone becomes a zone IV to our eyes. This means that if there is normal contrast in this scene there will be two stops of exposure difference between the indicated meter readings from each side of the face.

not skin tone is rendered too dark or too light. And, when skin tone is seen partly in light and partly in shadow (for example, a portrait taken by window light), the shadowed portion is seen as a zone IV. This previsualization is constant as long as we wish to create a normal appearance of skin tone in a photograph. Zone VI is skin tone as we normally see it, and zone IV will be the skin tone in shadow when compared to brightly lit skin.[2]

An experiment will confirm this. In even light, such as the shadow created by the north side of a building on a sunny day, take the palm of your hand and move it around (with your eyes squinting if necessary) until it is brighter than at any other point. This will happen when your palm is pointing directly at the main light source, in our example, the north sky. Take a meter reading from your palm and write it down. Next move your hand around until the palm appears darkest. This will be where the back of the hand blocks the main source of light and the palm is turned 180 degrees from where it was before. Take another meter reading.

A comparison of the two indicated meter readings should reveal a two-stop difference. This is the difference between zone IV and zone VI in a normally previsualized

---

[2]This is true for all Caucasian and light-skinned people. Black skin is another matter. Although the majority of black skin is previsualized as zone V, there is too much variation to make a general rule. Comparison with a known value such as a gray card will help establish a personal standard.

scene containing skin tones. If you do not have a two-stop difference, then you have made the readings in uneven light (either too directional or too contrasty) or in light that does not have enough intensity. Try another scene. North light between 10:00 A.M. and 3:00 P.M. on a sunny day will always produce a two-stop difference.

Contrast and Exposure Determination with Skin Tones

Once you have discovered a scene in which your hand-in-light/hand-in-shadow creates a two-stop difference in indicated meter readings, you have found a normal contrast scene. In it, tones can be previsualized the way your eye sees them. Then, without having to take any other meter readings, you can place exposure using the zone IV skin tone reading, and develop the film normally to print these tones on #2 grade enlarging paper. You can also use this method to judge the contrast of other scenes. A three-stop or more difference between the two readings (as will happen in direct sunlight) indicates that the contrast is high. Three stops difference indicates an $N-1$ scene, four stops $N-2$, etc. By the same standard, a one-stop difference indicates a scene of low contrast ($N+1$), and the need for longer-than-normal development. In each case, determine the expo-

sure through a zone IV placement of the reading from the palm of the hand in shadow.

The Short Cut

This method is the basis of our short-cut system. Rather than being limited to portraits, using skin tone previsualization to determine exposure and contrast can apply to any scene. Remember that once you use a shadow tone to place an exposure, all the other tones fall around it, either higher or lower, depending on whether they reflect more or less light. But, it isn't necessary that the tone metered for the exposure placement be in the scene as long as it is in the same light. Using the previsualization of skin in shadow as a zone IV and placing that means that all of the shadow tones in the scene will be recorded on the film as they relate to zone IV. Any tone that reflects the same amount of light as the skin in shadow will become a zone IV, any that reflects one stop less will become a zone III, and so on. As long as you previsualize and place the skin tone accurately, then the other tones fall accordingly.

Contrast determination works in a similar way. The range between the normal previsualization of skin in light and skin in shadow is adequate to determine contrast. One shadow and one highlight are the minimum needed. They indicate the contrast of any scene as long as the readings are made in the same

light as the scene being photographed, and as long as the previsualization is one that relates to normal skin values.

The Hand as a Standard Reference

If you apply the technique of previsualizing and metering skin tones to just the palm of your hand, the process becomes standardized and even more simplified. As you discovered from the experiment of locating the brightest and darkest points of a particular scene, the hand is flexible and easy to use as a target. The palm is large enough when held at a comfortable distance to fill the frame in a 35mm camera with a normal lens. Remember that the object in the viewfinder does not have to be in focus to take an accurate exposure reading.

With the hand as a standard reference for both highlight and shadow zones, the exposure and contrast readings will be accurate as long as you ensure that the light falling on the subject is the same as the light in which you make the meter readings. Whether the scene

Skin tones can also relate to a scene in which they don't even appear. The zones VI and IV of skin-in-light/skin-in-shadow have the same reflectance as any other zone VI and IV values in the scene if they are in the same light. Other tones in the scene will relate to skin tones too. Any that reflect less light than zone IV skin will be rendered in zones III or less. As well, the tones that reflect more light than zone VI skin will be in zones VII and higher. Using the hand as a substitute for metering can be a valuable technique when the previsualized tones such as the snow-covered lake and the forest cannot be metered by walking up to them.

is next to you, across the street, or a distant landscape, light will be reflected in the same relationship to the tones that you have metered.

Using the zone system this way has several advantages for the photographer. It means that you won't have to disturb a scene by walking up to meter it. It means that once you select a place to make photographs, you can make an exposure and contrast determination and not need to change it unless the light intensity changes. Finally, it means that if you are in a situation where you cannot physically approach a scene, you have substitute previsualized zones close at hand! (Other advantages have been pointed out: The palm of your hand is a piece of equipment that won't cost you any extra money, requires little maintenance, is easily portable, and, most importantly, doesn't complain or talk back.)

Of course, this method, like most short cuts, is a compromise. It would be more accurate in the long run to previsualize and meter more than two zones, and ones that are farther apart than IV and VI. But, for most situations where a photographer wants a scene to look the way the eye sees it, this method is quite reliable. Especially since it suits the needs of hand-held cameras with built-in light meters, it encourages the use of more thoughtful craft with 35mm photographers. The rest of this chapter is devoted to the actual working method for this system.

Previsualization Test

Knowing that the hand in shadow is a zone IV and the hand in light a zone VI gives us a method to test our previsualization. Any suspected shadow zone can be compared with the hand in shadow. If the indicated meter readings are the same, then the tone is a zone IV. If the shadow area reads one less stop of light than the hand, it is a zone III, and so on. The same method will work for highlights. Compare an indicated meter reading of the hand in the strongest ambient light with the indicated meter reading of the highlight to be tested. This will show what zone the suspected highlight will fall into under a normal previsualization. The test will work in all but the most contrasty and flat lighting situations and is a good method to use when learning to recognize zone equivalents to common objects, such as those listed in Chapter 2.

Putting Theory into Practice

Working with this system involves only a few steps. The first priority is making sure that you are standing in the same light as the subject you are photographing. You do not necessarily need to be close to your subject; sunlight, for example, is the same in intensity everywhere.

The second step is to position the palm of your hand so that it directly reflects the brightest light in the scene. It is important to realize that a light source does not have to be direct like the sun or a lamp, but can be an object that reflects a large percentage of light relative to other objects in the area. Be very careful to observe this, especially in low contrast situations where there usually is not a direct source of light. Fill the viewfinder of the camera without casting a shadow on the palm of your hand (this may mean holding the camera at a slight angle) and take a meter reading. The exposure indicated by the meter is your indicated meter reading for a previsualized zone VI. Remember that exposure or, better yet, write it down.

Next, rotate your hand 180 degrees so that it blocks the direction of the brightest light in the scene, causing the palm to be in shadow. This is a previsualized zone IV. Take a meter reading of the shadowed palm. One caution: when metering in sunlight, be careful that the hand completely shades the viewfinder, otherwise direct sunlight hitting the meter cell can temporarily blind it. Again, remember the reading or write it down.

These two readings are all the information you need. Take the indicated meter reading from the hand in shadow and place it in zone IV by stopping down one stop. This exposure setting will not change as long as the light remains constant. If it does change, or if you move into an area lit in a different way,

The zone VI reading is taken from the palm of the hand as it reflects the greatest amount of available light. The technique is simple: just hold the hand close enough to the camera to fill the viewfinder (it does not have to be in focus) and record the indicated meter reading. Be careful that the camera does not cast a shadow on the hand as it is being read.

another reading of the palm in shadow quickly gives you the new exposure.

A comparison of the two indicated meter readings provides contrast information. Two stops of difference indicates normal contrast; three stops, N−1; one stop, N+1. With roll film, you need to make only one contrast determination because all the frames are going to be developed together.

When the hand is held in such a way as to create the darkest shadow that can be found, it is a previsualized zone IV. The reading technique is again simple: hold the palm close enough to fill the viewfinder and take the reading. Be careful not to let direct sunlight strike the meter cell, or your meter could be temporarily "blinded."

Rules to Remember

This method simplifies working with the zone system, but you should keep certain things in mind or problems can quickly set in. Although the need to meter every scene is eliminated, you must become sensitive to changes in light intensity. The exposure will change, for example, when the sun goes behind a cloud, or you walk into the shadow of a building or under a tree. It is best to check the shadow reading from the palm of your hand frequently until you feel confident of your ability to sense exposure changes.

Since a roll of film can be developed only one way, development time must be chosen carefully. Contrast will often change in the middle of a roll, so it is important to establish certain rules. The first rule is to pick a development time that fits the majority of the frames on that roll. Contrast changes even less often than exposure when working with this system, so make note of it when it does occur, and plan development to fit as many of your frames as possible. This is a good argument for shooting 20-exposure rolls instead of 36, or even bulk loading shorter lengths, so that you can shoot an entire roll before you move on to a different contrast situation.

An alternative to the first rule is to develop your film for the contrast in the most important frames on the roll. Many people forget this seemingly obvious point.

Solving Special Problems

Even with these basic rules, special situations may occur. First, you may encounter a scene in which almost no difference appears in the meter readings between the hand in shadow and the hand in light, indicating that the contrast is very low. In such a case, place the skin tone in zone VI, and plan for normal development if possible. The negative will print well on high contrast paper.

If an N+1 scene must be shot on a roll that you have already decided to develop normally, simply forget the shadow reading. Instead, use the indicated meter reading for the hand in light and place it in zone VI. This negative will print well on a high contrast paper. If you should encounter an N−2 scene on a roll destined for normal development then again, forget the shadow reading and place the hand in light in zone VI. Use a low contrast paper when you print. These are not ideal solutions, as sometimes highlight detail will be lost in the print, but remember that this is a way to manage the problem of having many different contrast situations on a roll of film that can be given only one development.

Possibly the most difficult situation is an N+1 scene on a roll that must be developed for N−1. In this situation the skin tone in light should be placed on zone VI½ or VII (a stop and a half or two stops more than the indicated meter reading). When that frame is developed for N−1, the highlight densities will be about zone VI.

The most important rule of thumb is that all things being equal on a roll of film with two different contrasts, it is better to develop for the higher contrast scenes (i.e., the least amount of time). It is easier to add contrast in printing than to reduce it.

Using Paper Grades in Printing

Another way to handle the problem of multiple contrast scenes on a single roll of film is through the various paper grades. With correct placement of exposure, the shadow density of all the frames should be correct, even if they vary in contrast. The correctly developed frames will print as previsualized on normal contrast (#2) paper. You can then use other contrast grades to correct for the negatives that are too flat or too contrasty. Usually a change of one grade of paper equals a change of one zone in contrast. For example, an N+1 scene on a normally developed roll of film can be printed on a grade 3 paper. Variable contrast filters when used with the right papers function in much the same way as graded papers.

The following chart will suggest the various possibilities:

Development of Film	Contrast of Scene and Paper It Will Print on
N	N+1 on grade 3
	N−1 on grade 1
	N+2 on grade 4
N+1	N on grade 1
	N+2 on grade 3
	N−1 cannot be printed
N−1	N on grade 3
	N+1 on grade 4
	N−2 on grade 1

Although the above plan will produce good results, keep in mind that the best possible way to print is with a negative matched to a normal grade paper. Contrast grades other than #2 do not translate the tones in the negative accurately. The higher grades pull tones farther apart as a corrective for underdeveloped flat negatives. The lower grades push the tones closer together so that overdeveloped high contrast negatives will look normal. This distortion of tones is not quite the same as the tonal changes created by development.

Final Words of Caution

Experience reveals two common problems when getting used to the hand-in-light/hand-in-shadow method of metering. One problem is fully filling the frame with the palm of the hand so that the read-

Paper grades can correct many of the problems that are encountered when several different contrast scenes can be developed only one way. The most important consideration is to properly expose the negative so that the shadow tones are correct. The top photograph is an N +1 scene that was on a roll of film developed normally. It is printed on grade 3 paper. The bottom photograph is an N −1 scene that was developed normally and printed on grade 1 paper. In both, the tones appear as previsualized with a minimum of burning and dodging. Although using paper grades to correct for improper development is not an ideal solution (see text) it is a useful compromise for the roll film user.

ing is accurate. This can be a problem especially when metering the hand in shadow, as the light source bleeds around the hand and causes underexposure. The second problem is the opposite situation, getting so close with the camera that a shadow is cast on the hand. The result can be overexposure. Pay close attention to the distance between the positions of the hand and the camera.

Finally, it is important to realize that the only way to master the zone system is through practice. The techniques of previsualization, metering, and film development do not simply yield themselves on the first attempt. The tool of the zone system is most valuable when it is learned thoroughly enough to be used spontaneously. It has even been suggested that you have begun to master the zone system when you start asking people in what zone they would like their morning toast.[3]

---

[3]Minor White, *The Zone System Manual* (Hastings-on-Hudson, NY: Morgan & Morgan, Inc., 1968), p. 59.

9
The Zone System and Color

The Accuracy of Color

Many people have the mistaken impression that color prints and transparencies produce images that are more real than images in black and white. Quite apart from the difficulties of defining the term "real" we need to understand just how what we see is changed by rendering it as a color image. We tend to confuse the ability to see color in our images with the way we see the world itself. In fact, the added dimension of color often creates just another departure from an accurate view of the world. The many reasons for this have to do with the nature of the color materials themselves, and more importantly the nature of our language and our memory.

In the color process, the reproduction of colors depends on the response to light of multiple layers of silver emulsion and their subsequent replacement with manufactured dyes. Each formulation of a color emulsion renders color in a different way. Kodak's current Ektachrome process emphasizes cool colors (blues and greens) while Kodachrome tends to render warmer colors (reds and yellows) with more intensity. These generalizations are subject to change whenever the formulas are reconstructed, which seems to happen about every five years. Fujichrome, another popular transparency film, has a more pastel color rendition than Kodak's

products. Color prints have the same problem—they are made with dyes that cannot reproduce true colors, even when the actual colors are used as a reference in printing. In every case, the materials interpret the color scene. To work effectively in color, we must accept the distortions of the materials and learn to work with them.

Thinking in Color

Margaret Mead observed that Eskimos have 17 words for the color *white*, each one defining a particular snow condition. In the English language less than a dozen commonly used words describe color. This limited lexicon means that we have no clear concepts for more than a few colors, and this in turn limits our ability to see and distinguish color differences. We see symptoms of this problem in the convoluted adjectives used by interior decorators and car manufacturers in naming colors.

The lack of precision in our language often forces us to make vague generalizations about what we see. Josef Albers wrote that when a room full of people is asked to visualize the color *red*, everyone will think of different shades of that color. Even when a specific shade of red that is familiar to the audience is mentioned, such as "Coca-Cola" red, the results will be no different. Shown various shades of red and asked to pick out the one that

Previous page: At first appearance, color seems to add reality to a photographic image. In fact, the opposite is often true. This image employs a slow shutter speed and a swirl of red to create a fantasy of movement and color that relates only in the imagination to the reality of the bullfight. Previsualization of color is an important step in understanding how color affects what we photograph.

We have a very poor memory for color. Look at the shades of red here and try to decide which is closest to matching a familiar red color, such as that used in "Coca-Cola" advertising. Make a guess and then compare. Like most people, your guess will probably be wrong.

matches their memory of a "Coca-Cola" sign, invariably there will be a disagreement. Albers thought that this is due to the fact that the mind has a low retention of memory for color. Whereas sound patterns, like the beat of a popular song, can be remembered for long periods of time, the vision of color is fleeting.[1]

Describing Color

Every color has three significant features that define it: *hue, saturation,* and *luminance. Hue* is the name of a color, such as green or red. It refers to a specific wavelength or part of the color spectrum. *Saturation* (sometimes called "chroma") indicates the purity of a hue. A pastel color, which has a hue mixed with a lot of white, is desaturated. Strong, vivid hues are saturated. *Luminance,* also called "value," refers to the visual appearance of brightness in a color as it appears next to other colors in a scene. It is a subjective quality.

These terms can describe any

color in any situation. For example, red is a specific hue, the longest wavelength of visible light. Pink is a desaturated red, one that has a large amount of white mixed in it. On the other hand, Hemingway's description of blood in the bull-fighting ring indicates a very saturated red. If a red car is parked half in the shade of a tree and half in the sunlight, each half would have a different luminance but retain the same hue and saturation. Only if the hood of the red car reflects a "hot spot" of direct sunlight will that area be desaturated. Black is a hue with no saturation and very little luminance, while white also has no saturation, but a high luminance.

These concepts allow our language to deal with subtleties in color. They are equivalent to learning the scale of zones in black and white because they help us to translate what we see with our eyes into what the materials see. This is important because of the changing nature of color to our perceptions.

---

[1]Josef Albers' *Interaction of Color* (New Haven, CT: Yale University Press, revised edition, 1975) is one of the best texts for the color photographer wishing to learn about how color interacts with our perceptions.

Relativity of Color

In color, as in everything else, our perceptions are relative. A classic

*In this photograph of the hood
of a red car, we see illustrated
the different ways of describing
color. The car is a single hue, a
specific shade of red. The saturation of the red color remains constant except where
there is a glare of sunlight. This
desaturates the red. A high
luminance can be seen in the
red of the sunlit portion of the
car, and a lower luminance
where it is shaded. Understanding the terms* hue, saturation, *and* luminance, *and
using them to describe the colors we see, will help overcome
the language problem that
keeps us from seeing colors
accurately.*

demonstration of this kind of relativity is to place one hand in a bowl
of warm water and the other in a
bowl of cold water. After about 20
seconds, move both hands to a
third bowl containing lukewarm
water. The hand coming from the
bowl of warm water will then feel
cold, while the hand that had been
in the cold water will feel warm.
This kind of relativity is also present
in the way we perceive hue, saturation, and luminance. Simple experiments show how this happens.

Hue does not exist for us without
a reference. A room illuminated
only by blue light gradually appears
to us to be lit by dim, normal color
light. (A trick in the production of
night scenes for motion pictures is
to filter everything in blue.) Try looking at an even light source (such as a
light box or a blank white painted
wall) through a piece of blue ace-
tate. Gradually your eyes will stop
seeing color. If you introduce an object of a different color into your

field of vision, the blue reappears
along with the new color. Printers
learn to see color imbalances in
prints by glancing at the print
quickly and then at a standard reference tone rather than staring at
the print.

Saturation is seen in a relative
way too. For this experiment, several shades of the same color paper
are needed. Lay them out in strips
from the least saturated to the most
saturated so that they almost touch.
Next, cover the most saturated half
of the strips. From the ones that are
still visible, pick out the one that is
most saturated. Then cover all but
that strip and uncover the rest.
Comparing the strip you had
picked as being the most saturated
with the others now visible, decide
how your perception of that particular strip has changed. Repeat
this experiment with other colors,
until you reach an understanding
of how your perceptions of color
saturation can change.

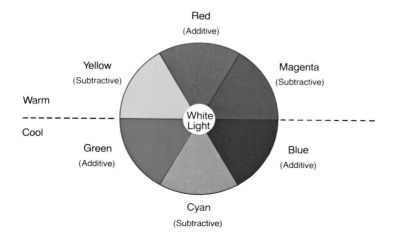

This "color wheel" shows the relationship of both the additive and subtractive primaries. In a color image, contrast is determined by these relationships. As well, there is a psychological contrast created by the difference between "warm" and "cool" colors.

Luminance can also be made to change visually. Take two squares of the same piece of color paper. Place one square in the middle of a larger piece of white paper, and the other on a larger piece of black paper. The color square on the black field will appear to have greater luminance than the one on the white field. You can also try this experiment with other colors substituting for the black and white backgrounds.

Color Contrast

ADDITIVE PRIMARIES
Our understanding of contrast must also adapt to color materials. We have defined contrast in black and white as the reflectance difference between the highlights and shadows. In color this difference is not as important as the contrast between the colors themselves. To understand this, one must first realize that white light is made up of three basic or "primary" colors: red, green, and blue. The relationship between each might be best imagined as an equilateral triangle in which each color takes a point, or

in the relationship indicated by the color wheel illustrated here. These colors are called "additive" primaries because each "adds" its color to the total of white light.

SUBTRACTIVE PRIMARIES
For every additive primary, there is a complementary color, one that contains all other colors except that additive primary. These complementary colors are cyan (a mixture of blue and green), magenta (equal parts of red and blue), and yellow (made up of red and green). Each subtracts its additive complement from white light and so is called a "subtractive" primary. Cyan is the complement of red, magenta the complement of green, and yellow the complement of blue. Their relationship to the additive primaries are shown in the color wheel.

CONTRAST OF PRIMARIES
Contrast in color images occurs when colors that are not next to each other on the color wheel appear next to each other in the image. A blue next to a green, or a yellow next to a cyan, will give the

appearance of great contrast, even when the two colors reflect the same amount of light. Partly this is caused by the structure of our eyes. Each color is a different wavelength of light, red being the longest and blue the shortest. The eye must

focus differently for each wavelength, and when two primaries are next to each other, the eye cannot resolve them both at the same time. They will seem to "vibrate" as the eye focuses back and forth for each color.

The exposure index test for transparencies (see p. 105) is based on the correct rendering of a textured highlight value. In this image, shot with Koda-chrome ASA 64 film, the white brick wall was previsualized as a zone VII. An indicated meter reading of the wall was then opened up two stops to place it in zone VII. Exposure indexes from E.I. 500 to E.I. 8 were tested. These three illustrations, evaluated by projecting them in a darkened room, show an underexposure (a), an overexposure (c), and a correctly exposed zone VII value (b).

(a) Underexposure

(b) Correct Exposure

(c) Overexposure

WARM AND COLD CONTRAST
Color also creates contrast between warm and cool colors. Red, yellow, and magenta are warm colors. We tend to associate them with warm feelings, such as those created by the light of the sun or a fire. Blue, green, and cyan are cool colors, which we associate with cool water, forests, and even ice. Whenever a cool color is placed next to a warm color the sense of opposite feelings that is created increases the contrast.

COLOR CONTRAST EXPERIMENT
To demonstrate how color contrast is different from black and white, try this experiment. Collect several pieces of color paper that each have a single solid color. Construction paper will work well. Use only pieces that are large enough to be read with your light meter. Find the two colors that to your eyes have the most contrast and place them side by side. Then compare indicated meter readings from each. If they are more than one stop apart (which is unlikely if you have chosen carefully), find two other pieces of color paper that still have a visual contrast, but have one stop or less difference in reflected light. When

you find these two colors, previsualize them as zones in a black-and-white image. With one stop or less difference, they will only be one zone or less apart. This would make a flat image. Yet, you have chosen them because as colors they have a great deal of contrast. Unlike black and white, any color image with these two hues will be high in contrast.

COLOR HARMONY
In color, a low contrast image is one in which the colors that predominate are next to each other on the color wheel. This harmony of colors does not necessarily depend on light reflectance. Harmonious colors could have large differences in the amount of reflected light and still not have as much visual contrast as opposite colors whose reflectance values are close together.

In working with color films, it is important to visualize the contrast as being as much a product of the colors present in the scene as it is the amount of light being reflected. High contrast is created by contrasting colors, low contrast by harmonious colors. The differences in light reflectance are much less important when using color materials than when using black-and-white.

Using Color Materials

TRANSPARENCIES
Transparencies are viewed by transmitted light, which means that the range of tones between highlights and shadows are not as much a matter of concern as with prints. Transmitted light passes through the image only once. Prints, on the other hand, are viewed with the light that is reflected off the base

(a) Underexposure

(b) Correct Exposure

(c) Overexposure

An exposure index test for color negative film (described on p. 106) is based on an examination of shadow values. The best way to run a test is to photograph the gray card and color/density scale pages of the Kodak Color Dataguide *at a variety of test indexes (see text for suggestions). If you examine the negatives with a magnifier, the correct exposure will be indicated when full separation of tones occurs in the reflection density scale. In these illustrations, (a) shows an underexposure. In it the density scale (the narrow vertical column) is blank and shows no separation in the bottom three segments. Illustration (b) indicates the correct exposure; all segments of the density scale are separate, distinct tones. Illustration (c) is typical of an overexposure. There is excessive density in all areas, but because the image is formed by layers of dyes rather than silver grains, the density will not necessarily prevent the making of a good print. When in doubt while shooting color negatives, it is always safer to overexpose.*

support material. As light bounces off the base, it must pass through the image in the emulsion twice, once going, and once coming. This doubles the visible density and limits what can be seen. To one used to looking at prints, a transparency offers a much greater range of tones and subtlety. This is fortunate, because unlike a negative, the tonal range of a transparency cannot be manipulated through development. To do so would cause serious changes in the reproduction of color (known as color balance). Thus, knowing the contrast of a scene in terms of light reflectance is not a crucial factor when working with transparencies.

Exposure is another matter. Careful exposure is necessary, as the film in the camera is also the final product. Neither underexposure nor overexposure can be corrected in a secondary step such as printing. The principles of correct exposure for transparencies are the same as for negatives. Areas of least

density in the image are the ones most controlled by exposure. In a positive image these areas are the highlights. This means metering off the important previsualized highlight in the scene and opening up the required number of stops for placement. All the other tones will fall relative to the placed zone. The darker values will be present no matter what their light reflectance is as long as the highlight is correct. This is possible because of the transparency's ability to render such a wide range of tones.

EXPOSURE INDEX TEST FOR TRANSPARENCIES

The exposure index for transparencies is based on a correct rendering of highlights. The most useful highlight for this test is zone VII, which has the least density and still shows recognizable detail. Zone VIII has less density, but is too close to clear film to be distinguished as a separate tone.

In this test, first find a standard zone VII value, such as a textured white sweater or a white painted brick wall. Set your meter to the index that you wish to test, take a meter reading, and place that reading in zone VII by opening up two f-stops or shutter speed equivalents. Development of the film should be strictly according to the manufacturer's specifications, regardless of the contrast of the scene. Test enough exposure indexes to allow a range of four stops less than the film's ASA and three

stops more. Evaluation of the images should be made by projecting them in a situation similar, if not identical, to the one in which your slides will usually be projected. Look for the image that best renders a zone VII value, full of texture, but not too dark, as an indication of the correct exposure index.

SHOOTING WITH COLOR TRANSPARENCIES

With the correct index, exposure involves previsualizing highlights only. Select the most important highlight, meter it, and place it in the previsualized zone. There is no need to measure the contrast range between highlights and shadows. Using your hand to determine exposure is even simpler: take the indicated meter reading from the palm of your hand in the brightest light and place it in zone VI by opening up one stop. As long as you are in the same light as your subject (as noted in Chapter 8) the exposure will be correct. The contrast of the image will be primarily a matter of the colors that you select to be in the scene.

COLOR NEGATIVES

All color negative films commonly used today are compatible with Kodak's color negative developing process. This is convenient and necessary since the correct balance of color dyes in a negative is difficult to attain. Industry standardization has been the answer. This does not mean that all color negative film has

Facing page: This image, shot on transparency film, combines many of the ideas of color theory and working methods mentioned in this book. First of all, it illustrates color contrast between the primaries of cyan, red, and green. As well, each color has different luminances where it is partially in the shade and partially in the sun. There is desaturation of color in the very brightly sunlit background seen through the crooked arm. To make this photograph, the technique of metering and placing the hand was used. Since the highlights in the scene were too small and moving too much to meter directly, this was the only practical method. The palm of the hand was located in a patch of sunlight (the same as the most important highlights) and the indicated meter reading was placed in zone VI.

the same response to color. Each emulsion has a different character, centering primarily around the saturation of certain colors. The best plan is to try the different brands available until you find one that suits your personal sense of color. Do not be awed by the word "professional" on the film label. In some cases the so-called amateur films produce better results and require less care in handling.

EXPOSURE INDEX TEST FOR COLOR NEGATIVES
Color negatives, like black-and-white negatives, must be exposed for the proper shadow density. Adequate detail in a zone III area indicates correct exposure. A simple exposure index test could be devised in which a zone III value is metered and placed using a variety of test indexes. An examination with a magnifying loupe to see which frame actually produces detail in the zone III area would reveal the proper index.

A more controlled test is to photograph the gray card and color/density scale page of the *Kodak Color Dataguide*. Starting with a test index that is eight times greater than the film's ASA, make a sequence of exposures that continues to an index eight times less. This will test seven different indexes and bracket the film's ASA by three stops. You can make meter readings from the gray card so that there is no need to change exposure for placement.

Process the film according to the manufacturer's data. Identify the index of each frame and place them all in 35mm slide mounts. Then project them in a sequence that begins with the highest index. Observe the image of the black-and-white density scale that is part of the page that you photographed. Each step is a separate reflection density from highlights to shadows. The correct index will be indicated by visible separation for all the steps. As you view the negatives starting from the highest index, you will see more and more of the density scale steps becoming separate tones. If ultimately you have difficulty in deciding between two frames of the test, always decide in favor of the frame with the most exposure. Unlike black-and-white, color negatives do not suffer any ill effects from overexposure. Overexposure of the negative by as much as two stops will not add greatly to the print exposure time, will not drastically affect the color balance, and will not add additional grain, since the final image is made up of dye layers rather than particles of silver. Overexposure is only insurance against loss of shadow detail.

CONTRAST CONTROL WITH COLOR NEGATIVES
Although contrast in color film is primarily a product of the colors in the scene being photographed, the overall range of light reflectance between previsualized highlights and shadows should not be completely forgotten. The contrast range of the original scene has some effect on

the final image, just as it does in black and white. If it is important to control this type of contrast in color negatives, you can modify the current Kodak C-41 development process. Although not recommended by Kodak, you can change development time with a minimum of effect on the balance of the color dyes and the ability to change the contrast by a one zone expansion or contraction. The times given are based on Kodak's recommended normal development time for the first roll of film processed in fresh developer.[2]

Contrast of Scene	Development Time
N−1	2 minutes 40 seconds
N	3 minutes 15 seconds
N+1	4 minutes

Kodak also makes a color printing paper designed to increase the contrast of a color negative. The paper's designation is "Ektacolor Type 78," and it is a substitute for the normal contrast "Ektacolor Type 74." Combined with an N+1 development, the total increase in contrast will be equivalent to N+2.

BLACK-AND-WHITE FILMS USING A DYE IMAGE

Now on the market are several types of black-and-white films known collectively as "silverless" films. These

[2]Development information is from *The Color Print Book* by Arnold Gassan (Rochester, NY: Light Impressions Corp., 1981).

are essentially color emulsions that have only one gray dye layer instead of several color ones. The major features of this type of film are that the dye-produced image shows no grain and the overall density is not as much affected by over- and underexposure as a comparable silver-produced image. However, the basis for light sensitivity of these films (as for all color emulsions) is silver, and the film is still subject to the controls of the zone system. Underexposure will cause a loss of shadow detail, and overexposure, while not as serious a problem as with standard black-and-white film, will eventually degrade the image.

Run the exposure index test for "silverless" films in the same manner as for color negative films. Development is with Kodak's C-41 process and contrast can be controlled through the development time modifications mentioned for it.

It should be pointed out that the silverless black-and-white films do not have permanent images. They will fade at approximately the same rate as any color emulsion.

Conclusion

Using the zone system with color materials is not that much different from using it with black-and-white. The ability to previsualize the scene and place the exposures are necessary for both. The major differences lie in the fact that contrast in color is determined more by the colors themselves than by light reflectance. Also, there is less flexibility in processing color films than with black-and-white because of the need to maintain color balance within the three layers of emulsion.

Ultimately, the most important aspects of the zone system are universal to all photographic materials and photographers. Much has been made in this book about learning the terminology of the zone system as one would learn another language. This involves not simply rote definitions but the ability to think in the new language. When you make the effort to learn, a totally new form of expression opens up to you. No longer must you default your creativity to the camera machinery and the chemical process. Now you can assert your control over the materials. Minor White was a photographer whose work with the zone system exemplified its creative potential. He called this use of it a "higher form of doing."[3] The freedom of choice given to you by the zone system makes you completely responsible for your photographic images.

---

[3] Minor White, *The Zone System Manual*, p. 98.

Index

Photo Credits

Page 1	Judy Canty	Page 45	Reid Callanan
Page 3	Eva Rubinstein	Page 56	David Torcoletti
Page 5	Craig Stevens	Page 57	David Torcoletti
Page 7	Judy Canty	Page 65	Bill DePalma
Page 11	M.P. Curtis	Page 75	Dan Bifere
Page 14	Nancy Roberts	Page 79	Judy Canty
Page 15	Kathleen Chmelewski	Page 80	Judy Canty
Page 16	Nancy Roberts	Page 85	Reid Callanan
Page 17	Kathleen Chmelewski	Page 87	Kathleen Chmelewski
Page 18	Kathleen Chmelewski	Page 89	Reid Callanan
Page 19	Ken Ross	Page 94	
Page 25	Reid Callanan	(top)	M.P. Curtis
Page 30	Judy Canty	Page 94	
Page 33	Judy Canty	(bottom)	Kathleen Chmelewski
Page 37	Judy Canty	Page 97	Terry Eiler
Page 39	Judy Canty	Page 100	Red car, courtesy of Paul
Page 40	Judy Canty		Raitz Sales and Service
Page 42	Barry Perlus	Page 104	Julie Oesterle